Fire & Ice

Books by David E. Fisher

FICTION

Crisis
A Fearful Symmetry
The Last Flying Tiger
Variation on a Theme
Katie's Terror
Hostage One

NONFICTION

Creation of the Universe
Creation of Atoms and Stars
The Ideas of Einstein
The Third Experiment
The Birth of the Earth
A Race on the Edge of Time
The Origin and Evolution of Our Own Particular Universe

Fire & Ice

THE GREENHOUSE EFFECT, OZONE DEPLETION, AND NUCLEAR WINTER

David E. Fisher

1817
HARPER & ROW, PUBLISHERS, New York
Grand Rapids, Philadelphia, St. Louis, San Francisco
London, Singapore, Sydney, Tokyo, Toronto

 For information address Harper & Row, Publishers, Inc., 10 East 53rd Street, New York, N.Y. 10022.

FIRST EDITION

Designer: Kim Llewellyn

Library of Congress Cataloging-in-Publication Data
Fisher, David E., 1932–
Fire and ice: the greenhouse effect, ozone depletion, and nuclear winter/David Fisher.
p. cm.
ISBN 0-06-016214-7
1. Air—Pollution. 2. Greenhouse effect, Atmospheric. 3. Ozone layer depletion. 4. Nuclear winter. I. Title.
TD883.1.F57 1990
363.73'92—dc20 89-45652

90 91 92 93 94 CC/HC 10 9 8 7 6 5 4 3 2 1

This book is for the Maine gang:

Bill, San, Liz and Y, Stash, Gogs, M, Hoss, Miss,
Scotty, and Toots.

Some say the world will end in fire,
Some say in ice. . . .

—Robert Frost

Contents

PART ONE: THE PROBLEMS

PART TWO: THE SOLUTIONS

PART ONE

The Problems

1

Fire

What's the point? I mean, what is the bloody point?
—John Cleese, as Basil Fawlty

In 1908, analyzing the spectrum of light from Comet Morehouse, astronomers detected the presence of cyanogen, a deadly gas. In 1910 Halley's comet was due to return. A newspaper story mentioned that Halley's tail was known to extend about thirty million miles out into space as it swept by the sun. Since the tail is blown out from the comet's head by the pressure of sunlight, it always extends away from the sun, that is, toward the earth as it swings around the sun. Another newspaper story mentioned that at closest approach the comet would be within fourteen million miles of Earth. Simultaneously throughout the country several reporters managed the necessary subtraction and correlation, and the next day their newspapers reported that Earth would pass through the comet's tail, which was composed of deadly cyanogen.

Astronomers were consulted. They admitted that it was true, but that the cyanogen was only a minor component of the tail and that the tail itself was so tenuous that no one on earth would possibly be able to tell when we passed through it. There would be no effect.

Perhaps not, but there was a pre-effect: hysteria. A popular song looked at the bright side:

No more politicians,
No more tariff schemes,
No more trust conditions,
No more quick-rich dreams!

Bang! Annihilation!
Smash! We fly to bits!
There's some consolation
If the comet hits!

Scientists reassured everyone that the comet was not going to hit, it would never come closer than fourteen million miles. But when asked about the comet's tail they could not give such a simple, unequivocal answer. Yes, they admitted, we *would* pass through its tail, but— Well, yes, there *was* cyanogen in the tail, but—

Hysteria. In Kentucky churchgoers flocked to services to "meet their doom." In Pennsylvania coal miners refused to enter the pits, afraid of being trapped down there; in Colorado coal miners wouldn't come out of the pits, feeling safer hidden there. In Texas lumberjacks fled to revival meetings, closing down operations. In New York, Pittsburgh, Denver, and San Francisco people committed suicide. In Oklahoma the Secret Followers gathered to appease the comet by cutting out the heart of a sixteen-year-old virgin, in response to their leader's conversation with God, who told him that otherwise "the world would end and the heavens would be rolled up like a scroll following contact with the tail of the Comet."

In Texas two salesmen went through the territory ahead of the comet, selling comet pills and gas masks. They were arrested, but subsequently freed when an angry mob gathered outside the courthouse and threatened to lynch, not the salesmen who had swindled them, but the police if they didn't let the men free so they could continue to distribute their pills and masks.

They fared better, as it turned out, than the citizens of Towaco, New Jersey, who flocked to the top of Waukhaw Mountain to watch for the comet after being offered cold cash by two "scientists" for their descriptions. They stayed up there all night, and returned to their homes in the morning to find all their chicken coops empty and the "scientists" long gone.

The most imaginative end of the world was forecast by someone who knew something—but not very much—about chemistry. The nitrogen in cyanogen (C_2N_2) would combine with the oxygen in the earth's atmosphere, he predicted, to form nitrous oxide (N_2O), or laughing gas. It would kill everyone on earth, but they would "dance, deliriously happy, to an anesthetic death."

It might not be a bad way to go. Certainly it would be better than the end people feared when, on Halloween Eve, 1938, they tuned in their radios to hear that the Martians had landed and were systematically exterminating the people of Grovers Mill, New Jersey. Once again people panicked, but once again the world did not end.

Stories of catastrophes have always fascinated us. They have come down to us as tales of Götterdämmerung and Noah's flood, as tales of alien invasions and cosmic cataclysms, and somehow we're always interested. Today, however, there is a difference. When Comet Halley approached the earth it was the scientists who were calling for calm, who were telling everyone that there was no danger. Today it is the scientists who are sounding the alarm. The modern twist lies not in the nonsense about cometary vapors or in the Wellesian fantasies of warring Martians, but in the discovery that catastrophes on a global scale are not mere imaginings but actual fact. They have happened in the past, they will happen again.

In 1822 Gideon Mantell was a country doctor practicing in the Brighton area of Sussex, on the southern coast of England. One day in the spring of that year Dr. Mantell made a house call accompanied by his wife, Mary Ann. While waiting outside the patient's house, walking up and down, she saw what appeared to be a large tooth embedded in a rock on the edge of the lane. When Dr. Mantell finished his visit and came out, she showed it to him. An amateur paleontologist, he was fascinated; he chipped off the rock fragment containing the tooth and took it home.

There had been previous stories about giant bones found in the region, but if these were given any credence at all they were thought to be the remains of human ancestors larger than the human species is today. Everyone knew, after all, the story of David and Goliath, which "proved" that giants had at one time existed. Dr. Mantell recognized this particular tooth, however, as being reptilian rather than human. It closely resembled the teeth of the iguana, and so in an 1825 paper that served to get him elected a Fellow of the Royal Society he named the creature *Iguanodon* and suggested that it was a giant lizard.

A reinterpretation of the previously found giant bones followed, and eventually they were recognized as belonging to a new type of creature, one that had never been observed by man. Richard Owen, a British paleontologist working in the second half of the nineteenth century, named them *dinosaurs.* Since then we have been studying their teeth and bones and footprints, trying to learn more about how they lived and how they died.

Particularly how they died. Anyone who has ever seen a life-size reconstruction of a Brontosaurus or Tyrannosaurus Rex must wonder how these monsters were eliminated from the face of the earth. For a hundred years this has loomed as a Mystery of Science, with various explanations offered and almost unanimously rejected, until in 1980 the father and son scientists Walter and Louis Alvarez found that a thin world-wide layer of iridium-enriched sediment had been deposited on the earth's surface at just the time of the dinosaurs' demise. Since iridium is an element scarce on the surface of the earth but relatively abundant in meteorites, they suggested that a giant meteorite, or perhaps an asteroid or comet, had smashed into the earth and killed the dinosaurs by altering the climate, either heating it suddenly by energy deposition or cooling it suddenly and/or cutting off photosynthesis by raising a cloud of smoke and dust.

The details are far from settled, but the consensus is clear that *something* happened, some cataclysmic catastrophe such as an asteroidal impact or perhaps a series of terrible volcanic eruptions, and through these our climate and atmosphere were suddenly changed to a point where not only the lords of creation, the dinosaurs, died *in toto* but so did about 75 percent of all species living at the time.

Such events are fascinating to wonder about, and so the asteroid/volcano scenario and the Time of the Great Killing have made network television and the cover of *Time,* and have been featured in newspaper headlines and comics and in cartoons in magazines as diverse as the *New Yorker* and *Penthouse,* such as the one by Gary Larson *(The Far Side)* which shows dinosaurs slyly smoking cigarettes, with the caption "The real reason the dinosaurs became extinct."

The catastrophe which destroyed the dinosaurs and three-quarters of all other living things happened sixty-five million years ago. There are three circumstances today, brought on by our own activities, which threaten to bring down upon us a series of cataclysms no less terrible. The first of these is the most familiar, a holocaust that has become almost hackneyed in the telling and retelling of repeated warnings.

The alarm was sounded on November 10, 1932, when Stanley

Baldwin, past and future Prime Minister of Great Britain and at the time Lord President of the Council, brought public recognition of a new age of warfare to the world. He addressed the House of Commons and simultaneously broadcast to the people at home, using the technologic marvel of radio for the first time to hammer home an appalling message. "I think it is as well for the man in the street to realize," he warned, "that no power on earth can protect him from being bombed. Whatever people may tell him, *the bomber will always get through.*"

And indeed he was right. In the autumn of 1940 London was blitzed by an average of two hundred aircraft a night for sixty-one consecutive nights until finally the tables turned. On July 27, 1943, nearly a thousand British bombers dropped over two thousand tons of bombs on Hamburg, most of them incendiaries, turning that city into a burning, melting quagmire of horror. The temperature reached one thousand degrees in the center of town, igniting the world's first firestorm. The superheated air rose so fast it sucked in outside air in the form of hurricane-strength winds which force-fed the fire still further and blew helpless people like leaves into the burning center of destruction where they actually melted into pools of burning fat. On the outskirts of the storm other people were stuck in molten asphalt, suffocating and igniting. More than 40,000 people died that night. In the early spring of 1945 the American Twentieth Air Force topped the RAF's record by burning Tokyo, starting a conflagration that totaled sixteen square miles of intensely populated city, killing more than 80,000 people. A few months later the atomic bomb was dropped, and within another decade we were seriously contemplating the possibility of destroying ourselves.

In 1953 our plan in case of war was to hit Russia with 900 nuclear bombs, a number we were assured was enough to reduce Russia within two hours to a smoking, radiating ruin. But that wasn't enough: by the next year we had 1,500 bombs and by the middle 1960s we and the Russians had thousands of bombs aimed at each other. Today there are about 50,000 Russian and American nuclear warheads waiting to fly out of silos, submarines, and bombers; enough to incinerate in hydrogen-fused fireballs every city on earth many times over.

As the nuclear arsenal inexorably grew during these decades we turned to each other and asked in bewilderment, "What's the point of all these bombs? What is the bloody point?"

The only answer, Kurt Vonnegut told us with a shrug, was simple: "So it goes," he said.

But he was wrong. It gets worse.

2

Ice

The ice ages may return due to our altering the global climate. We are talking about the end of the world—doomsday—in 25 years.

—*New Times,* March 1975, quoting various experts

There are many things we do not know about this world. For instance, in 1927 a small disk-shaped piece of iron-nickel metal was found buried twelve inches deep in a wheatfield in Washington County, Colorado. This was strange, for aside from the fact that there are no iron deposits in this region, metal of that composition should oxidize and form at least a coating of rust, but the Washington County chunk was rust free. The large content of nickel indicated a similarity to meteorites, but other characteristic meteoritic features were absent.

In 1959 I held a postdoctoral position in the chemistry department of Brookhaven National Laboratory, and together with my supervisor, Oliver Schaeffer, I measured the helium-3 content of that piece of metal. Helium-3 is a minor isotope of helium that is produced in meteorites by the action of cosmic rays, but that is largely absent from terrestrial materials. This particular piece of iron-nickel had abundant helium-3, and so the small mystery was cleared up: it was a chunk of metal that had just happened to fall from the sky into Washington County.

At the beginning of the nineteenth century there were a series of similar but more striking mysteries. In the forests of Yosemite National Park there stand isolated boulders among the trees, some of them as big as houses. They sit on the ground, strangers in a strange land, not dug into the earth as if they had dropped from the sky but rather as if they had been placed there gently by a giant hand.

Small deposits of dirt, soil, and even pebbles might be carried along by torrential floods and deposited far from their places of natural origin, but no water could carry boulders as large as these. Recent studies indicate no helium-3. Similar boulders, together with smaller rocks, are found occasionally dotting the fertile soil landscapes of Kansas and Illinois. Chemical and mineralogical studies indicate a close similarity—indeed an identicality—to rock outcrops many hundreds of miles to the north. Such exotic rocks are not unique: in England there are occasional boulders clearly identical to rocks in the mountains of Norway.

In New York City's Central Park, some of the rocks on which children play exhibit long, straight striations while others have polished surfaces, as if a giant hand had scoured them with Brillo. In Europe as well as in North America large rock deposits are found beyond the openings of valleys, as if someone had taken a gigantic pail of boulders and sent them careening and banging down the valley and out onto the plains.

In 1821 a Swiss engineer, Ignaz Venetz, suggested that all such world-wide anomalies could have a single cause. High in the Swiss Alps the mountain valleys are filled year round with snow-covered ice, which flows and expands down the valleys in the winter, gouging rock and soil out of the land over which it scrapes, polishing and grooving the bedrock, finally halting and beginning to melt in the spring. At the base of each such glacier flow, when it melts and retreats every year, is deposited and left behind a mass of ice-rafted pebbles, dirt, and occasionally even large boulders. Venetz suggested that the giant boulders found in Yosemite, Kansas, and Illinois, as well as in Europe, had been carried there by enormous glaciers, which at one time covered vast areas of the northern hemisphere. The polished, striated rocks of Central Park bear witness to the passage of the scraping ice.

In 1836 another Swiss scientist, Louis Agassiz, set out to disprove this theory, which he considered frivolous, based on too little evidence. As he studied the glaciers and their movements in the Swiss Alps, however, he became convinced that glacial movements could indeed account for the widespread observations; finally he became convinced that there was no other explanation. He and the generation of scientists who followed him were able to prove that a number of features and deposits that today lie beyond the range of any known ice sheets were in fact produced by prehistoric ice movements during the past two million years. Evidence of well-developed zones of surface weathering and soil deposits mixed in with the glacial features showed that there had been not one but at least four different time periods during which the land was covered with ice thousands of feet thick. It had polished the rock, gouged out striations, and pried loose giant boulders, carrying them along many hundreds of miles before depositing them when eventually the ice melted and retreated once again to the frozen North.

These time periods were called the ice ages. Down to a latitude including most of the current midwestern grain-producing states and New York, nearly half of Europe, and on across Siberia and Asia the land was covered with slowly flowing sheets of ice that were miles thick. They ground flat the forests and fields beneath them, wiping out all life, carrying along their booty of gouged-out rocks and boulders, scouring the land with their weight.

Till the middle of the twentieth century we thought that there had been just four such ice periods in the past 2 million years. The idea was that somehow, roughly every 500,000 years or so, something went wrong with the earth's climate or the sun's heating mechanism, and temperatures plunged below freezing for half the northern hemisphere.

In 1946 Harold Urey was a professor of chemistry at the University of Chicago. Born in Indiana in 1893 and raised there through high school, he taught in country schools in that state and in Montana before attending Montana State University. In the 1920s he earned a Ph.D. at Berkeley, and in 1931, while at Columbia University, he discovered deuterium, the heavy isotope of normal hydrogen. Three years later he won the Nobel Prize for that work. During the Second World War he was one of the inner circle of scientists who produced the atomic bomb, and one of the most vociferously opposed to actually using it. After the war he settled in Chicago to pursue more peaceful interests. Together with a student, Stanley Miller, he planned and carried out an experiment that revolutionized our concept of the origin of life. They mixed together gaseous solutions of the simple chemicals water, methane, ammonia, and hydrogen, sparked them with electricity, and found that they had been transformed into amino acids, the molecules of which life is composed. Our ideas today of the origin of life on earth—in which a primitive atmosphere composed of just such simple chemical species is supposed to have been shocked by lightning, volcanic eruptions, and cosmic rays to produce the amino acids—are based on this experiment. Later, as professor at the University of California, Urey was to become the driving force behind our space experiments, turning the military race to the moon into a valid scientific enterprise.

In December of 1946 he was in Switzerland, giving a lecture at the Eidgenoschische Technische Hochschule on the results of his experiments with the isotopes of carbon and oxygen. His theme was that though the isotopes of an element are chemically identical they can still behave differently under certain conditions. He claimed, for example, that when water evaporates the three isotopes of oxygen—which differ in their mass according to their atomic weights of 16, 17, and 18—will evaporate at different rates, the lighter 0-16 atom evaporating more readily than the heavier 0-18. Therefore water distilled into steam will be lighter than the residue left behind, and the water vapor in clouds must be lighter than oceanic water.

A crystallographer in the audience, Paul Niggli, pointed out that if all this were true it would follow that fresh and sea water must have different oxygen isotopic ratios, and one could perhaps measure these ratios in carbonate deposits of unknown origin and determine if they had precipitated from lakes or oceans.

Urey thought about this on the trip back to Chicago, and in carrying out the detailed calculations involved he found that the process would be dependent on the temperature of the water involved. And suddenly a light clicked on. As he later said, "I suddenly found myself with a geologic thermometer in my hands."

Well, not quite in his hands; at that time it was only in his head. It took another four years to put together a mass spectrometer capable of detecting the extremely small isotopic differences that would correspond to a temperature difference of a few degrees centigrade, but by 1950 he and his students were ready to begin measurements.

Cesare Emiliani at that time had just received a Ph.D. under Urey's direction. (In fact it was his second Ph.D. degree; Emiliani had grown up in Italy during the war years and earned his first Ph.D. at the University of Bologna in 1945.) Coming to Chicago for further study he picked up the second degree, and then began work on

Urey's new geothermometer. At the beginning of his work the university was visited by Hans Pettersson, leader of the Swedish Deep-Sea Expedition of 1947, one of the first postwar explorations of the ocean floor. Dr. Pettersson convinced Urey and Emiliani to concentrate on the carbonate skeletons of the tiny one-celled marine animals called foraminifera. Some species of this animal live on the ocean bottom, and when they die their skeletons accumulate there to form part of the sediment that covers the ocean floor. Pettersson supplied the Chicago laboratory with numerous samples brought up from the bottom by "piston coring," in which a hollow cylindrical tube is dropped into the sediment, and the entrapped material is hauled back up to the waiting ship.

The line of research begun then has continued up to the present day, and has revolutionized our understanding both of the causes of the ice ages and of their chronology. We now know, for example, that the ice ages were not occasional perturbations to the earth's normal temperate climate, but are in fact the more usual condition. Temperatures as warm as those in which we live today are the exception, and the studies taught us, further, that such *interglacial* periods do not last terribly long, geologically speaking; the earth usually cools off again and slides back into an ice age after about ten thousand years.

Ten thousand years is, of course, a long time on human time scales, but as the chronology of successive ice ages began to be developed from the deep-sea sediment studies, a disturbing development appeared: Not only do the warm interglacials last only about ten thousand years, but the present interglacial period—during which all of our civilization evolved—began just about ten thousand years ago. The implication was clear: we might be about to slide down into a new ice age. And so scientists began to take a closer look at the historical record of recent global temperatures, and what they saw did not reassure them at all.

From the late 1930s into the early 1970s data taken by several meteorological groups indicated unambiguously that the world was cooling down, with average global temperature dropping by nearly a full degree centigrade over that interval. If that doesn't sound like much, one has only to realize that the average global temperature during the ice ages was only a few degrees centigrade cooler than today. Putting the observed temperature decline together with the chronological oxygen-isotope data, scientists the world over began to think that perhaps indeed we had already begun our descent into the next ice age.

And the descent was coming with an important difference. Previous declines into glacial periods took place relatively slowly, over roughly thousands of years. But there was now the additional factor of man's influence on his world, an influence that appeared to be hastening the decline into hundreds of years, or perhaps even decades. Evidence was beginning to accumulate that we could no longer treat the earth's atmosphere and oceans as an infinite reservoir that would go its own way despite whatever small perturbations we added to it. Instead it was becoming clear that mankind's activities were now of such a scale as to have a clear and definite effect; we were beginning to overwhelm God's septic tank with our own excrement.

Two things in particular were worrying. We were adding to the atmosphere increasing amounts of a chemical known as chlorofluorocarbons, compounds of chlorine, fluorine, and carbon, which destroy the ozone layer. We'll discuss this in more detail later; for now the important point seemed to be that the ozone layer, sitting at the base of the stratosphere and absorbing a significant amount of solar energy, was an important fulcrum about which the jet stream and thus the world's weather balanced. If it were destroyed, or significantly depleted, atmospheric warming would be affected, the path of the jet streams would be deflected, and the worry was that the overall effect might be to cool the earth and hasten its decline into the next ice age.

Even more worrisome were the increasing amounts of smoke and soot being thrown into the atmosphere by our always-increasing use of fossil fuels. Such debris floating around in the air would deflect sunlight from the ground and thus make the earth cooler, acting in effect like dark, wintry clouds. And so in 1972 an international conference was held at Brown University to study the question "The Present Interglacial, How and When Will It End?"

The attendees agreed on a general model of the earth's climate and weather, and so too should we.

3

Climate

The price in human suffering of continued ignorance of the causes of climate change may already have become unacceptably high.

—U.S. National Research Council, 1974

We live in a universe governed by inviolate laws; that is our salvation and our tragedy. We do not understand all these laws; that is what turns our tragedy into farce.

One of these laws is the second law of thermodynamics, which specifies that heat must always flow from hot to cold. The earth is an isolated body whirling through space, surrounded by temperatures of close to 0° absolute, or 273° below zero centigrade. It has internal sources of energy, particularly the radioactive isotopes of uranium, thorium, and potassium within its crust and mantle, which provide the heat to generate volcanoes and plate tectonics, but such energy sources contribute only a minuscule fraction of the heat that warms the surface and atmosphere of our planet. The earth would be a frozen ball if it were not for the radiant heat of the sun.

The sun's energy is fed by gravitational contraction, which maintains a thermonuclear fusion furnace in its core. In this process the sun's hydrogen, comprising nearly 90 percent of its mass, is being continually transformed into helium with a simultaneous loss of mass; the lost mass, according to $E = mc^2$, is turned into energy, which is convected up to the surface and there radiated out into space.

All hot bodies radiate electromagnetic energy. We see this continually on earth: strike a match and it glows brightly because the fire is hot. Turn on an electric stove and soon the coils begin to glow visibly. In an incandescent light bulb electrons are forced through a thin wire, the friction of their motion heats the wire, and thus it radiates light.

All hot bodies radiate, but not at the same wavelength. Hotter bodies radiate at shorter wavelengths, cooler bodies at longer wavelengths. Light, visible light, is what we call that part of the electromagnetic radiation with wavelengths of several hundred nanometers.* When radiation at these wavelengths impinges upon our eyes it sends chemical signals to the brain; this is the process we call "seeing." Our eyes can do even more; they can differentiate between electromagnetic radiation of different wavelengths (within a small spread), and the recognition of such wavelength differences gives rise to what we call "color." Red light, for example, has a wavelength of approximately 700 nanometers; blue or violet light, at the shorter end of the spectrum, has a wavelength of 450, and the other colors lie between these extremes. Our eyes cannot detect radiation with wavelengths either longer than red or shorter than blue, but such radiations exist and we are continually surrounded by them. Our radios and television sets are instruments capable of detecting radiation with wavelengths in the meter range and turning them into sight and sound. We turn these instruments on and we hear and see whatever is being broadcast. Obviously the radios are not creating the radio waves, but are simply receiving those that are there filling the room all the time, whether the set is turned on or not.

The sun, because of its surface temperature of about 6000° C., happens to radiate most of its energy in the visible region of the spectrum, as shown in Figure 1. (Actually, it doesn't just "happen" to radiate visible light. It's probably more true to turn the argument around and say that our eyes evolved a sensitivity to this region of the spectrum precisely because that is where the sun sends out most of its energy. Stars with surface temperatures different from that of the sun emit most of their radiation at different wavelengths, and thus there are radio, ultraviolet, and even x-ray stars. If we had evolved on some other planet circling one of these stars, we probably would have evolved eyes sensitive to those wavelengths. Even on this earth, some nocturnal hunting snakes have eyes that are sensitive only to wavelengths in the infrared region; such wavelengths are emitted by warm living bodies, and so the snakes can "see" their prey in the dark.)

*A nanometer is one billionth of a meter.

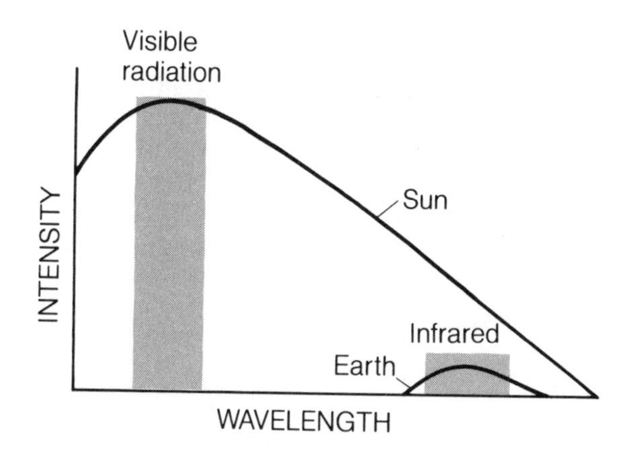

Figure 1

The earth's average surface temperature is about 15°C. instead of the sun's 6000°, and so the earth emits much less energy and at a much longer wavelength, again as shown in Figure 1. But though the earth is cooler than the sun, it is still much warmer than the surrounding regions of space, and so it too must emit radiation. Thus the earth is warmed by the radiant heat of the sun and in turn radiates heat out into space. The balance between the sun's incoming radiation and the earth's outgoing radiation determines the surface temperature and the climate of the earth. This balance is governed by several factors, one of which is the ratio between the earth's reflectivity and its absorption of sunlight.

If the earth were covered with a spherical mirror nearly all the sun's radiation would bounce off and be reflected back into space, with only a minute amount being absorbed by the mirror. The earth, shielded by the mirror, would be cold.

To a very large extent, we have such a mirror. The ice and snow that cap the regions around the North and South Poles are exceedingly bright in sunshine because they reflect nearly all the sun's light. A plowed dirt field, on the other hand, appears dark because it absorbs most of the sunlight, reflecting very little. If the earth were covered with dirt fields it would absorb more of the sun's energy than today's earth does, and temperatures would be much hotter. If the polar ice caps were to grow and encircle the earth, they would reflect away more heat and the earth would become colder.

The reflectivity, or *albedo* of the earth, is one of the two main factors affecting our climate. The other, even more important, is the atmosphere, which covers us like a loving blanket.

Our heat comes from the sun, and so, as we move higher into the atmosphere and therefore closer to the sun, things should get warmer. But observations tell us precisely the opposite. As altitude increases from sea level to thirty or forty thousand feet, the temperature drops sharply.

It isn't a function of science to insist that our theories are correct (that is the domain of religion); rather, science's whole *raison d'être* is to examine the world as it is, seek out those points where the observations do not tally with theoretical expectations, and then investigate the causes of the discrepancies. That's how we make progress—not when we're right, but when we're wrong. In the case of the vertical heat distribution of the atmosphere, the explanation of the discrepancies becomes immediately obvious: the air is *not* heated by the sun.

Air, as we are all aware, is transparent to visible light; if it were not, we couldn't see through it. This means that the majority of the sun's radiation, which after all lies in the visible part of the spectrum, passes right through it. If it passes through the atmosphere, it can't heat it; a substance is heated by electromagnetic radiation only when the radiation is absorbed and has its energy transformed through atomic motion into heat. To illustrate this, imagine that you are holding a magnifying glass and concentrating the sun's rays upon a blade of grass. The grass is heated, and will burst into flame. If a pane of plain glass is inserted between the magnifying glass and the grass, the sun's rays will go right through it since glass is obviously transparent to visible light. The heating of the grass will not be impaired, and the glass pane will not get hot. On the other hand if something that is not transparent to light, such as a piece of paper, is placed between the two, it will absorb the sun's rays and will itself be heated while the grass behind it remains cool.

So the sun irradiates the earth, and most of the sun's energy passes directly through the atmosphere without heating it, and is either reflected or absorbed by the earth. The reflected radiant energy simply bounces off again, passing through the atmosphere on its way out as easily as it did on its way in, and is lost to space, affecting the earth not one whit. The absorbed energy, however, heats the earth. As shown back in Figure 1 the earth must radiate this energy out again, but because it is at a lower temperature than the sun the radiation emitted from the earth will be at a longer wavelength. In fact, most of the radiation will come off in the infrared region of the spectrum, at a wavelength of about 50,000 nanometers.

There are several components of the earth's atmosphere that are *not* transparent to wavelengths of this magnitude; primary among these are carbon dioxide and water. These molecules absorb infrared radiation, and in doing so cause the atmosphere to be heated up. In this manner the radiant energy of the sun eventually warms our air—but from the bottom up, so that the air closer to the ground is warmer just as toes held closer to a fire are warmer than fingers held further away.

The atmospheric phenomenon of allowing the sun's radiant energy to pass through and then trapping the earth's re-irradiated energy is known as the "greenhouse effect," an idea first coined in 1827 by the French mathematician Jean Baptiste Fourier, who compared it to the mechanism by which a greenhouse can maintain warm temperatures even in a northern winter. The glass, like the atmosphere, allows the sun's radiant energy to penetrate but traps the internally-emitted infrared radiation and keeps it inside. In fact, it is this effect that is responsible for making our planet habitable.

Our neighboring planets in the solar system, Venus and Mars, are inhospitable to life. Venus is too hot, Mars too cold. In a qualitative sense this is what we would expect simply from their positions relative to the sun, since Venus is closer and Mars further away than the earth. But quantitatively the numbers don't jibe. Venus is three-quarters of the earth's distance from the sun, Mars one and a half times as far away. Because the energy gathered by the planets varies with the square of the distance, Venus should absorb twice as much energy, and Mars half as much, as does the earth. But the temperature of an absorbing body, according to a well-documented law first discovered by Josef Stefan and proved by Ludwig Boltzmann, varies only slightly with incident energy (approximately as the fourth root of the energy), so that the temperature variations of these three planets should be only slightly different; say, within a hundred degrees of each other. Yet Venus has a surface temperature of over 400°C., while Mars is 60° below zero.

The reason for the difference lies in the greenhouse effect. Venus has nearly five hundred times more carbon dioxide in its atmosphere than earth has, and this traps the outgoing radiation efficiently and heats the planet. Mars has less than one percent of the earth's gases, and so doesn't retain its surface heat. The earth's carbon dioxide budget lies between these two, and is just right for the evolution and continuance of life. But it doesn't come with a permanent guarantee that it will stay that way.

4

The March of Ice

The characteristic shortness of previous warm interglacial intervals indicates that they reflect a precarious environmental balance. This raises the possibility that man's interference with climate may trigger some environmental chain reaction and speed the natural trend to a new ice age.

—"The Present Interglacial," Scientific conference held at Brown University, 1972

At one time it had been thought that, in order to change this rather temperate world into an ice-covered, frozen ball, a catastrophic plunge in world temperatures would be necessary. This belief made it impossible to understand the causes of glaciation, for as the isotopic temperature measurements later showed, at the height of the last ice age—when half the northern hemisphere was covered with icy glaciers more than a mile thick—the average global temperature was only 5°C. colder than it is today. Obviously the cause did not have to be a total cataclysm such as the turning-off of the sun, but could have been a more subtle triggering mechanism.

One had already been proposed but neglected by the scientific community, both because its effect had been thought to be too small to induce such a large change and because it would lead to a rapid succession of effects for which there was no evidence. But the Urey-Emiliani temperature measurements changed all that.

It is difficult but not impossible to go back into the past. We do it with memory and with records and stories, but if we are honest we realize that travel distorts the scenery: the things we remember, the things we read about and write about, often didn't happen quite the way they are remembered and discussed.

Geologists travel to the past by digging. One of the cardinal principles of the discipline is the law of deposition, which makes the reasonable claim that sediments are laid down from the bottom up.

The Grand Canyon, for example, is a sedimentary sequence: looking at it you see layer upon layer of sediments that were deposited at a time when the region was under water, and obviously the bottom sediments were deposited first, with later sediments falling on top of the earlier. So, as a geologist digs down into the earth, he is going further and further back in time.

But of course he has to pick the right places to dig. If the earth had been formed as a single sedimentary rock, beginning with a speck that is now the core, one could go back all the way to the beginning by digging down to the core. Aside from the impossibility of digging such a hole, that isn't the way the earth formed. Most of the earth's interior is composed not of sedimentary rock but of a fluid, continually convecting and ever-changing mantle. Large portions of the crust, however, have at one time been under water, and on these areas sediment forms and builds up: airborne dust falls onto the water and crud from rivers is washed into it and swimming and floating creatures die in it, and all these things slowly fall to the bottom and form a layer of sediment. As time continues so too does the process, and slowly the bottoms of all water reservoirs are covered with thickening sediments.

If conditions at the bottom are left undisturbed, the pressure of the overlying sediments and the heat of radioactive elements within them begins to cook the bottom layers into hard sedimentary rocks. Millennia later, sea levels may fall or the press of rising magma from beneath may lift the rock layers to dry land, as happened with the Grand Canyon, exposing the record of the past to present-day geologists.

The problem with such records is that they are incomplete. A sedimentary deposit may form and then be uplifted or exposed, and at that point the sedimentary process stops. Airborne dust that is blown onto the Grand Canyon rocks today, for example, is just as rapidly blown away again: the sediment no longer accumulates. At some later time the deposit may fall again below sea level (or may be covered by rising seas) and the process will begin again. A latter-day geologist, examining the rocks when once again they are exposed on dry land, would not see a continuous record of geological history: the time period corresponding to when the rocks were on dry land would be missing.

All exposed sedimentary deposits, if they are of sufficient age to be interesting, suffer from this fault since the history of our world is a succession of rising and falling sea levels. A misunderstanding of this fact, incidentally, is what led to the myth of the Missing Link. Because the sedimentary record is necessarily incomplete, so too must be the fossil record. We do not see a complete and slowly evolving record of evolutionary changes, not because there is a Missing Link between any two species, but because the "tape" on which the "music" has been recorded has been cut up and patched time and time again.

To gain a more complete and undisturbed fossil record, geologists have turned to the sea floor. The Atlantic Ocean, for example, was begun about two hundred million years ago when the ancient continent of Pangaea rifted apart to form the Americas on one side and Europe-Africa on the other. Since that time the ocean has been continually widening as mantle magmas surge out from beneath the Mid-Atlantic Ridge and push the continents apart, and never in that time has the deep ocean floor been raised above sea level.

Sediments cored from the bottom of the sea, therefore, have the most continuous records yet found, but even they are not complete. Burrowing animals at the ocean-sediment interface redistribute the sediment, pushing surface material deep down and bringing up deeper material, and sudden currents can wash great masses of sediment away to be deposited in a slurry somewhere else. The coring process itself must disturb the sediment, and can easily lose the top or bottom portions.

A variety of techniques are now available to allow us to tell when such losses have occurred, if not to tell precisely what material was lost. Use of radioactive dating and geomagnetic polarity-reversal techniques allows the correlation of one core with another, and gradually over the years a comprehensive and reasonably continuous record of the isotopic temperature changes in foraminifera has been built up, providing an accurate global climatic history. By the mid-1970s the results showed a definite periodicity to the onset of the successive ice ages, corresponding to the cycles predicted by the old discarded theory of Milutin Milankovitch.

This theory was based on the fact that the earth's motion around the sun is not a perfect circle, nor even a perfect ellipse. Instead there are periodic oscillations around a generalized curve, due to the gravitational influences of Jupiter and Saturn in addition to that of the sun. Three particular variations are in the shape, or eccentricity, of the orbit; in the obliquity, or the angle that the north-south axis of the earth makes with the plane of the earth's orbit; and in a wobble of the axis, called precession. We can discuss this most easily by starting with the simplest, ideal case, in which none of these variations exists, and then proceeding to introduce the complications by a series of models. This will also serve to introduce the concept of model building with both its strengths and limitations, which is at the heart of the nuclear winter, greenhouse, and ozone controversies today.

Our object is to explain the earth's climate, both present and past. We start with a simple model, in which the earth revolves in a perfect and continual circle around the sun, at the same time spinning about its north-south axis, which is perpendicular to the plane of its orbit. This model reproduces in a qualitative fashion two of the three most important characteristics of the earth's climate, and so must be judged a reasonable success.

As the earth spins on its axis the surface area lit by the sun continually revolves around the globe, and so the model predicts and accounts for the regular alternation of night and day. It also tells us that the earth will be warmer near the equator than at the poles, as illustrated in Figure 2, where a square unit of earth's surface (a) at the equator receives all the radiation indicated by the volume *A*, while the same unit of surface area near the pole, because of its tilt, receives a smaller amount of solar radiation (indicated by *B*). Therefore the earth's surface receives more radiant energy per unit area near the equator than it does near the poles and, at least in this model where all other things are equal, must be warmer. Furthermore, since the surface of the earth curves in a regular way, the diminution of temperature from equator to pole will be steady, and this is roughly what we see: a gradually increasing coldness as we journey north and south from the equator.

The model's great failure is in the third major characteristic of our weather: the seasons. In this model the earth is always at the same distance from the sun, and every day and every night is twelve hours long, and no region of the earth experiences periods of warmer and colder weather (except for the night-day fluctuation). There is no summer, no winter on this world.

To account for the seasons in our model, we have to leave this

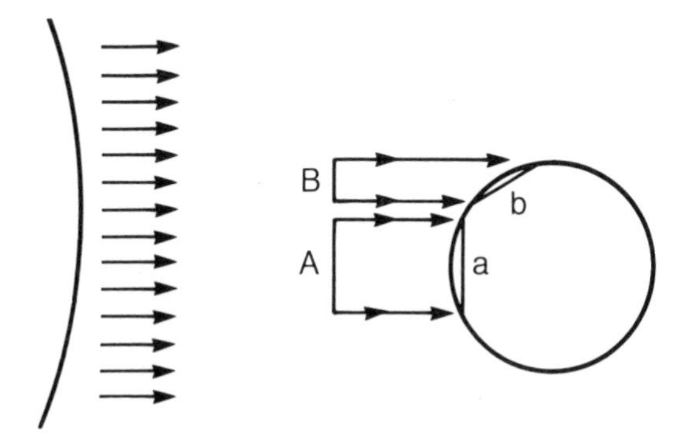

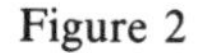
Figure 2

simple, ideal case, and begin to introduce complications. This is a common characteristic of all model building; one starts with the simplest case, which is often rather far removed from reality (but must have at least some relationship to it in order to be at all valid). The virtue of such simple models is not their reality but that they are readily understandable; they are a starting point. We then proceed to introduce complications in an effort to approximate the real world more closely.

We might at this point introduce the fact that the earth's orbit is an ellipse rather than a circle, and so the earth is sometimes closer to the sun and sometimes further away. Winter would then occur at the far point of the ellipse and summer when the earth is closest to the sun. But this model does not work: it would predict *global* winters and summers, which is not what the observations tell us. Northern hemispheric summers actually correspond to southern winters, and this fact is not explained by the model. So we discard it. In fact the eccentricity of the earth's elliptical orbit has only a minor effect on the seasons.

Let us instead address the fact that the earth's north-south axis is not aligned perpendicularly to the plane of its orbit, but instead inclines at an angle of 23.5 degrees. When the earth's North Pole is tilted toward the sun the Northern Hemisphere will receive more sunlight in a twenty-four-hour period than the southern, and these conditions will be reversed when the earth travels halfway around its orbit (six months later) and the South Pole then tilts toward the sun.

In the former case the Northern Hemisphere will experience days longer than its nights, as can be visualized easily considering the extreme case: if the north pole were to point directly toward the sun (as it nearly does for the planet Uranus) the northern hemisphere would be in continual sunlight during all twenty-four hours of the earth's daily rotation, and the southern hemisphere would be in unrelieved darkness. Longer hours of sunshine obviously mean more sunlight falling on the region during any twenty-four-hour period, but in addition the tilt puts more sunlight on the northern surface during each second of irradiation. Thus we have hemispherically opposed summer and winter, each occurring once during the yearly rotation of the earth around the sun, with the gradual change in axial tilt leading gradually from winter to spring to summer, and then to autumn and winter once again. The effect of the eccentricity of the orbit is that the earth happens to be further away from the sun during northern winter and closer during northern summer; this emphasizes the primary variation and makes northern summers and winters more extreme than those in the southern hemisphere.

This is a satisfactory model of today's climate system, but it would predict that the climate should have remained the same throughout geologic time. In order to explain the ice ages, further complications must be added. Seventy years ago Milutin Milankovitch, a short, stocky forty-one-year-old Serbian engineer, made the attempt.

Milankovitch was born in Dalj, Slavonia, a small farming community in what is now Yugoslavia. He earned a doctorate in technical sciences and later joined the faculty of the University of Belgrade. His first paper on the astronomical motion of the earth in relation to the ice ages appeared in 1920, followed by a series of papers and books extending over the next forty-odd years as he wrestled with the problem. In 1982 his son recalled an incident from that interval.

> Father and I were holidaying in Austria. One morning I jumped into a swimming pool and fractured my collar bone. I received pain-killing injections and was put in a temporary auxiliary splint made out of knitting wool.
>
> Later that afternoon and throughout dinner, father was very absent-minded. It was after ten o'clock when he finally

settled me into bed in my room, connected to his large room where there was a writing desk. He left the door ajar in case I needed him. Instead of going to bed, he went to his desk, pulled some large sheets of paper from the centre drawer, and sat down.

The effect of my injection wore off and the splint was cutting painfully into my shoulders and I could not sleep. He took his glasses off and looked at me. He was not seeing me—he looked through me. He sat down and wrote again.

The woolen strap was hurting me. I called out for him to lift me and put some cotton wool under the strap, but I could see that he did not hear me.

He kept on writing, and then he stopped and looked at the paper. He began to revise all that he had written down, talking to himself. Then he took another sheet of paper and started writing again. Slowly the tension of his face gradually dispersed and his usual calm expression returned.

He looked up and said, "Sorry, I forgot about you. Do you need anything?" He lifted me into a more comfortable position, put pads of cotton wool behind the strap, and said, "I think I've got it."

"You got what?" I asked.

"The differential equation covering the movement of the poles. It's eluded me for quite some time. But don't worry, I'm all right now."

He patted me on my shoulder, went into his room, and a couple of minutes later I could hear him gently snoring.

Milankovitch's theory dealt with the known variations of the earth's orbit around the sun, which occur on time scales of from 26,000 to 105,000 years. In the model so far described the tilt of the earth's axis is always pointed in the same direction, but in fact it wobbles (or *precesses*). In addition, the angle the axis makes with the imaginary line drawn perpendicular to the plane of the orbit is not always 23.5 degrees as it is today, but varies between 22 and 24 degrees. Since the tilt causes the seasons, it follows that the greater the tilt the greater the seasonal effects, that is, the colder the winters and the hotter the summers. This effect is magnified when the precession places the tilt direction directly toward the sun in summer, and is minimized when the tilt direction is toward the sun in winter. Finally, Milankovitch added in the effect of periodic changes in the orbit's eccentricity (or elongation of the ellipse). Because the periods are not in any simple numerical relationship with each other, and because the effects sometimes cancel and sometimes reinforce each other, the calculation of the periodicity to be expected in the earth's climate is not a simple matter, and even today it is not precisely agreed upon by all the people studying it. Clearly, however, it predicted a periodicity somewhere in the region of 26,000 to 105,000 years, which did not correspond to the facts as then known.

It was commonly thought at the time that there had been four ice ages during the past 2 million years. The average time between them was about 500,000 years, which did not fit the Milankovitch prediction, so it wasn't taken seriously by many people. But the isotopic work begun by Urey and Emiliani changed our ideas about the frequency of the ice ages, and today it is clear to most scientists that the Milankovitch variations combine to control the forcing of glacial periods roughly every thousand centuries.

One problem still remains, however: calculations indicate that the temperature changes induced by the orbital variations are not enough to plunge the world into icy conditions. This is particularly true because these variations do not change at all the total amount of sunlight received by the earth; all they do is change the distribution of sunlight over the globe. If the Milankovitch theory is to work, somehow the orbital variations must act as a trigger, setting off positive feedback mechanisms to magnify the effect.

There are a variety of possible feedback mechanisms, and it is not yet clear whether any or all of them are important in causing an ice age to begin. Two of the most important have been already discussed, the albedo and the greenhouse effect.

The albedo effect is particularly easy to understand. For simplicity we'll discuss only the northern hemisphere. The North Pole is capped with snow and ice, and permanent glaciers extend over much of Canada, Greenland, and Asia, as well as in mountainous regions such as the Alps. This snow and ice has an albedo of nearly 100 percent, reflecting away the sun's radiant energy that would otherwise be absorbed by the land or water beneath them.

Every winter more snow falls in the high latitudes and in the mountains, and the glaciers increase in areal extent, creeping always downward and southward. Every summer the tips of the glaciers melt and they shrink back to their previous borders. If the orbital variations make the summers less warm, less of the glaciers will melt, and the following winter they will start their yearly expansion from a base already expanded. If the next summer is again cool, the process will repeat and the glaciers will year by year grow and expand.

As they expand, they reflect away more sunlight, further cooling the earth, leading to cooler summers, which lead to increased glacier expansion, which leads to increasing albedo and cooler summers, and away we go.

The greenhouse effect is a bit more complicated but acts in much the same way. Somehow the orbital variations may induce a change in oceanic circulation or oceanic biological patterns, the combined effect of which is to transfer some of the carbon dioxide dissolved in ocean waters to deeper layers—or at least we know from detailed measurements that during the last glacial period the amounts of carbon dioxide dissolved in intermediate layers of the ocean decreased, and the amounts dissolved in the deep oceans increased dramatically. Dissolved carbon dioxide makes the water more acidic, which in turn makes it a more effective dissolving agent. If the bottom ocean waters are more acidic, they will dissolve the carbonate sediments. This shifts the balance of equilibrium among dissolved carbonate species, and within a few years makes the surface ocean waters capable of dissolving even more carbon dioxide from the atmosphere. This in turns leads to a lessening of the atmospheric carbon dioxide, which decreases the greenhouse effect, which lowers atmospheric temperatures and allows more snow and ice to fall and accumulate, and once again away we go.

The exact details of these positive-feedback mechanisms are not completely known. For example, we know from actual measurements that the carbon dioxide of bottom ocean waters was higher in the last glacial, and that this increase preceded the peak fall in surface temperatures, but we don't know exactly how the orbital variations induce this change in carbon dioxide oceanic distributions. The important point, so far as predicting the oncoming of another ice age is concerned, is that there are a variety of poorly understood feedbacks that can plunge us irreversibly into another ice age with only the triggering of a subtle onset mechanism. By the 1970s we were becoming aware that our own effects on this planet have not been entirely negligible, and the fear began to grow that perhaps without knowing it we might be triggering some such mechanism.

5

The End of the World: Part 1

The whole earth is held hostage. This is a time for ideas, not gadgets.

—Sergei Kapitsa, senior scientist at the Physical Problems Institute of the Soviet Academy of Sciences

When the last ice age began and the northern and mountain glaciers started their irresistible descent, plants and animals began to move down the mountainsides to lower regions, and from northern lands to the south. It was like a stream of refugees, except that it took generations and centuries instead of days and weeks. According to the chronology laid down in the deep-sea sediments, the coldest period of the last ice age occurred about 18,000 years ago, by which time a thick glacier covered nearly all of Canada and much of the United States down to the latitude of New York City, Chicago, and Seattle. On the other side of the Atlantic, ice covered everything down to Copenhagen, Berlin, and Leningrad.

In North America the glacier stretched from the Rocky Mountains to the Atlantic, rising abruptly at its southern edge like a white mountain stretching five thousand feet high, then sloping upward another five thousand feet to the north. Under that ice all life was crushed; at its edge all life was mangled. Toward the south the continent gradually warmed, turning into vast regions of tundra frozen in winter and stagnantly wet in summer, the home of hordes of hungry mosquitoes and roving herds of woolly mammoths. Further south, where the tundra was modified into spruce forests, the mastodons lived, together with moose, elk, giant beavers weighing five hundred pounds, and all sorts of rodents.

Man did not exist in America then. He probably entered the continent about ten thousand years ago, on the heels of the retreating ice, moving in from Asia across the frozen Strait of Bering, killing the fauna as he moved southward with the warming tide. Slowly the earth entered the present warm interglacial period, and our civilization began to flourish. But today the normal ten-thousand-year interglacial period is about up, and we have to wonder how our civilization might fare in another ice age.

Is there reason for concern? After all, we did manage to exist in the last ice age. Though a definition of the precise beginnings of modern man is impossible, it is clear that *Homo sapiens* has existed for at least several tens of thousands of years, a time period that extends right through the time of the great glaciers. If we survived that ice age as savages, could we not survive the next one as a sophisticated civilization?

Probably not. As savages we were free and nomadic; as a civilization of nations we are constrained and immobile. When the great ice sheets began their inexorable sweep southward twenty-some thousand years ago, the human tribe as well as lower orders of animals moved ahead of them, fleeing the encroaching ice. Following their food supplies, they managed to survive and even flourish, and when the ice retreated they followed it into North America. The whole world, after all, was never entirely crushed beneath the ice, nor were temperatures everywhere below freezing.

But what would happen today? If Russia became uninhabitable, where would the Russians go? If Germany were covered with ice, would the Germans be welcomed into Italy? If all of Canada and half the United States lay beneath miles of ice, would Nicaragua open its arms to receive the refugees?

Not according to our past record. Global war would surely ensue, and with nuclear weapons at our fingertips could extinction be far behind?

Even without the threat of our self-destruction, a new ice age would surely be a calamity dwarfing any in recorded history. (It is probably true that the advent of the last ice age hastened our evolution into an efficient society by separating the strong from the weak and putting survival pressure on the strong. So, to be realistic, we today must be grateful that our ancestors had to go through that ordeal. But enough is enough. Hitler argued that war is necessary because it purifies a nation's spirit in the cold flame of survival, but it would take a man equally as mad to claim that we would now benefit from another ice age.) And so when we saw that the average global temperature had fallen nearly 1° C. from the 1930s till the 1970s, we began to sweat profusely and coldly.

One degree centigrade doesn't seem like much, but the record of the onset of glacial conditions in previous epoches indicates that it previously took hundreds of years for such a temperature drop, and so we began to wonder what was happening. Instead of sliding gently into that cold night, had we already begun a precipitous slide down a slope made more slippery by our own excrement?

The closer we looked, the more that appeared possible.

The atmosphere consists of the *troposphere,* extending from the ground up to about thirty or forty thousand feet, and then the *stratosphere,* which extends out into space. The troposphere is the "normal" part of the air, warmed from the bottom by the earth's infrared radiation, with temperatures decreasing as altitude increases. The warmer air at the bottom expands and rises, and its space is taken by cooler, denser air falling in from above; this continual circulation accounts for our winds, and produces a well-mixed ocean of air. As the warm air rises higher and cools enough, its water vapor condenses into liquid and falls back down as rain. To sum up, everything we think of as normal weather occurs in the troposphere.

The stratosphere-troposphere boundary is marked by a layer of ozone, a gas that absorbs the sun's energy and heats the air above it so that the temperature no longer decreases with increasing altitude. Because of this, the normal circulation patterns of the troposphere do not occur in the stratosphere; in particular, convective mixing and condensation of water vapor into rain do not occur. This can have very severe repercussions.

In late autumn of 1815 the gigantic volcano Tambora on the Indonesian island of Sumbawa erupted, dwarfing most other eruptions in human history, and 1816 became known as "the year without a summer." Dust from the eruption was flung high into the stratosphere, higher than the tropospheric regions where continual vertical circulation of the air and the formation of rain would sweep it out, and so it floated there for months before slowly settling down. During that time it was blown around the globe, and everywhere it went the sun was sure to fade.

Acting as a thin veil, the dust blocked the sunshine and sent it ricocheting off into space, shielding the land below from both light and heat. It wasn't thick enough to turn daylight into darkness, but it was thick enough. American farmers and ranchers in the western territory looked out on fields dried and frozen by a changed climate and moaned in desperation. Though the average temperature was only about 1° C. colder than usual, severe effects followed. Snow fell in New England in June, the Vermont July was compared to a normal November, the mountains remained covered with snow, and frosts occurred more than once in every month of that lost summer, killing the crops. Wheat harvests were poor in both Europe and America, and in Brittany the wine grapes died. In Zürich the people began to eat their cats. There were food riots in Ireland, Wales, and France; there were armed revolutions in many of the market towns of Europe, with government troops firing into the mobs screaming for bread. Famine in India so weakened the population that the nation's most severe cholera epidemic broke out, and some medical historians have traced its spread from Bengal to New York. The one good thing that came out of Tambora was *Frankenstein.* In July Mary Shelley was in Switzerland with her half-brother and Lord Byron: "It proved a wet, ungenial summer, and incessant rain often confined us for days to the house. . . . 'We will each write a ghost story,' said Lord Byron . . ."

The cause of the lost summer was not immediately recognized. Some suggested it might be Ben Franklin's fault: The earth, they reasoned, was warmed by heat from the interior that was brought to the surface by electrical conductance. Lightning rods, which Franklin had invented and which many people were using that summer, must have upset the natural flow of electricity, thus chilling the earth. (In a similar manner, I remember the cool Florida winter of 1958 being blamed on the rockets launched from Cape Canaveral: a letter to the local Gainesville newspaper suggested that the rockets blasted holes in the atmosphere and "let in all the cold air from space." Don't laugh. This is in a country where according to a recent poll half the people don't believe in evolution and fully 88 percent think astrology is based on science. But maybe they don't know what astrology is. Once while attending a conference on Miami Beach I took a cab to a restaurant. The cab driver asked if I was in town with a convention and I told him yes, a conference on cosmology. He said, "Yeah? Hey, I got a cousin from Brooklyn who's a cosmologist!" I was afraid not to believe him; after all, Richard Feynman was somebody's cousin from Brooklyn. It turned out that his cousin sold cosmetics. So if the distinction between cosmetics and cosmology is blurred, perhaps it's understandable that we don't know the difference between astrology and astronomy.)

In fact, Ben Franklin did have a hand in the 1815 catastrophe: some thirty years earlier he had suggested that variations in the atmospheric dust load might absorb sunlight and cause cool weather on earth. But though newspapers reported both the bad weather and the observations of volcanic dust settling all over the world, no one seems to have made the connection.

Tambora was not the first volcano to upset the world's climate, nor will it be the last. (Mount St. Helens in 1980 had only a local effect because it erupted sidewards instead of vertically, and so its dust was not blown high enough to escape local rain; Krakatoa in 1883 produced only one quarter as much atmospheric dust as Tambora and so didn't affect much more than the beauty of sunsets.) But more frightening than a summer lost through violent but natural events like volcanoes is the threat of a longer-lasting, more severe blockage of sunlight due to our own activities.

Like Mount St. Helens, normal human activities do not throw dust high enough to reach the stratosphere, where it can get beyond the downward grip of daily weather processes. But there is another, more sinister difference between human activities and volcanoes: human activity is an ongoing thing. It isn't necessary that the dust thrown into the atmosphere stay there a long time, because even as it settles out we are throwing up more and more. The rate of dust settling will be proportional to the amount of dust injected, and so a steady-state situation results, with the deposition of dust from the air being matched by our production. The question is, what is the level of dust concentration in the air reached by this steady state? Does it result in enough atmospheric dust to block the sunlight and bring on another ice age?

In 1972 the answer to this question was far from clear, but the scientists who gathered at Brown University for a conference on The Present Interglacial: How and When Will It End? were worried.

In the summer of 1988 the newspapers were full of stories about heat and drought, and the coming of the greenhouse effect. In 1972 they carried stories about ice and snow, and the coming of the next ice age. In their opening and closing remarks to the Brown University conference, the conveners pointed out the evidence that was worrying the world and that had brought the scientists together for discussions.

The worries had begun with recognition of the global cooling that started in the 1940s, but there was increasing evidence that the situation was rapidly getting worse. The past winter, in particular, had been characterized by unusual cold and snow in America. The snow had even continued into springtime in areas that normally saw flowers instead of slush by April. Nor was it only in America that the climate seemed to be worsening. The winter had been unusually cold in Central Asia, and snowbanks at the time of the conference covered Baffin Island in the Arctic straits, which had been seasonally snow-free for the past century. Pack ice around Iceland was becoming a threat to navigation. Warmth-loving animals such as armadillos, which had expanded northward into the American Midwest in the first half of the century, were now reported to be retreating southward. All in all, the pattern was clear: it had been getting colder for half a century, and the most recent years were the coldest.

Several investigators pointed out the further dangers caused by man's activities. Comparing the present temperature patterns with previous interglacials, as recorded in the sediment record, they showed that the present interglacial is in its final phase and that if nature is allowed to run its course unaltered by man, events similar to those that ended the last interglacial should be expected within the next few centuries. But they went further, arguing that the characteristic shortness of previous warm interglacial intervals indicated that such intervals reflect a precarious environmental balance, raising the possibility that man's interference with climate at this point in postglacial time might become particularly risky. "Man's activities," they concluded, "may trigger some environmental chain reaction and speed the natural trend to a new cold phase of climate."

This concern was merely a reflection of the mood of the conference. One scientist from Stockholm argued that the present interglacial will "certainly be followed by the Future Ice Age," although further information was necessary before we could tell when. A group from Copenhagen, cooperating with the United States Cold Regions Research and Engineering Laboratory at Hanover, New

Hampshire, reported on their measurements of oxygen isotopic ratios in ice cores taken at Camp Century, Greenland. There had been a spectacular drop in the ratio, indicative of rapid cooling and the onset of the last ice age, which might have been triggered by accidental climatic events. "If so," they warned, "the conditions for a catastrophic event are present today [that might] trigger a full glaciation." They concluded by wondering, "Is man's present activity equivalent to such an accidental event?"

Another scientist, from the Hebrew University in Jerusalem, pointed out that changes in oceanic salinity might trigger a sudden glaciation. Stephen Schneider, of the National Center for Atmospheric Research in Boulder, said that the temperature decrease in polar regions during the past decade had been several times larger than the global average decrease, and that increases in particle concentrations (dust caused by human activities) could have a significant effect on climate. Past climatic variations, he said, had purely natural causes, but "we have recently realized that man has altered the face of the Earth and the composition of the atmosphere on such a large scale that this might modify these conclusions."

Nearly lost among all these fears of the next ice age was the voice of J. Murray Mitchell, Jr., of the National Oceanic and Atmospheric Administration (NOAA), who pointed out that as man burnt fossil fuels he not only threw dust and smoke into the atmosphere but also carbon dioxide, which would increase the greenhouse effect and counter the cooling induced by the particulates. He concluded that the greenhouse effect would overwhelm the dust, and that "the net impact of human activities on the climate of future decades is quite likely to be one of warming," thereby putting off the next ice age instead of bringing it on.

But this concern had been addressed previously. Professors Bryson and Wendland of the University of Wisconsin had argued in 1968 that "since 1940, the effect of the rapid rise of atmospheric turbidity [dustiness, due to pollution] appears to have exceeded the effect of rising carbon dioxide, resulting in a rapid downward trend of temperature. There is no indication that these trends will be reversed, and there is some reason to believe that man-made pollution will have an increased effect in the future."

By 1974 Stephen Schneider, writing with his colleague W. Kellogg in *Science,* reported that clouds are an important and heretofore uninvestigated contributor to the climate, through their role as the dominant reflectors of solar energy. The synergism of man and nature manifests itself through the clouds. Suppose that the heat we supply to the environment by burning fossil fuels more than compensates for the sunlight we bounce off by our dust. Then the atmosphere would get warmer. But the overall effect might be that the increased warmth would result in an increased evaporation of water, which would cause more clouds and reflect away more solar heat, thus cooling the earth.

The popular press and the nation as a whole picked up the general consensus that a freezing century was on its way. *Newsweek,* in an article titled "The Cooling World," warned in 1975 that there were "ominous signs that the earth's weather patterns have begun to change dramatically . . . with serious implications for just about every nation on earth. . . . The earth's climate seems to be cooling down," they said, noting the steady decline of global temperatures during the past half century, and quoting a Columbia University scientist whose satellite photos indicated a sudden large increase in Northern Hemisphere snow cover in the winter of 1971–72, and a study by two NOAA scientists showing that the amount of sunshine reaching the ground in the continental United States diminished by 1.3 percent between 1964 and 1972. All the data indicated, they reported, "that the present decline has taken the planet about a sixth of the way toward the Ice Age."

What to do? Suggestions began to pour in. Some scientists suggested that we cover the arctic ice cap with dark soot. This would absorb sunlight instead of allowing the ice to reflect it, thus bringing to earth more of the sun's energy and warming the atmosphere. Other scientists pointed out that this might melt the ice cap, raising the sea level hundreds of feet world-wide and flooding London, New York, Tokyo, Singapore, Rome, and nearly all centers of civilization. The Russians suggested that they might alter the flow of the Arctic rivers, and thus perhaps diminish the feeding of the ice cap to stop its growth.

Luckily, before any of these suggestions even began to be implemented, further studies showed that indeed we were in trouble—but of an entirely different kind.

6

UV-B

Quick, Henry! The Flit!
—Advertisement for insecticide, circa 1935

In late 1945 my father took me to an air show in Miami. The hit of the afternoon was a demonstration of the Lockheed P-80 Shooting Star, the Air Corps's new jet fighter. It appeared suddenly in the distance as a small spot and quickly materialized into a shiny white airplane hurtling toward us, eerily silent. With its nose tipped down it could almost reach the speed of sound, and so it very nearly caught up to the monstrous roar of its jet engine and flashed in front of us barely a moment after its crashing noise cracked against our ears. Trailing its rumble behind like the tail of a tiger, it clawed its way back up into the sky and was gone among the clouds, and I still remember the audible sigh of awe that went up from the crowd.

The next year Geoffrey Raoul de Havilland, chief test pilot for de Havilland Aircraft and son of the firm's founder, took their new Swallow jet up to try to break the world's speed record, which at that time was held by Britain's first operational jet fighter, the Gloster Meteor, with a speed of 615.8 miles per hour, just under the speed of sound.

The Swallow was a modification of the Vampire, de Havilland's new fighter that was replacing the Meteor, and there was no question that it was a faster airplane. In fact it was so fast that people were beginning to worry about what would happen if it came up against the "sound barrier." His Majesty's Government had actually banned any attempt to fly faster than sound, deciding to use radio-controlled models to investigate the possibility because of the general consensus that there was a wall in the sky at the speed of sound, a wall that would destroy any plane that came up against it. The government decision did not stop de Havilland, who was flying the company's own civil modification of the sweptwing tailless aircraft, nor did the fear of anything in the sky affect his actions. He had been the first pilot to fly the company's famous Mosquito wooden bomber, though the "experts" had predicted it would fall apart under the stress, and he had been the first to fly the Vampire.

In the late afternoon of September 27, 1946, he took the Swallow up for the record attempt and swept out of sight down the Thames estuary. He planned to open the plane up to its full speed at low level, then come back to the aerodrome and fly over the prescribed course for the record. But he never did.

Shortly after they lost sight of him at Hatfield aerodrome, they lost radio contact. There was no further word until the telephone rang nearly an hour later. An RAF officer and his wife, walking on Canvey Island, had heard the characteristic whistle of a jet. Looking up, they saw the strange tailless aircraft zooming along at high speed—and then they saw it suddenly disintegrate "as if it had run head-on into a wall in the sky." And so in fact it had: it had come up against the sound barrier. Ten days later Geoffrey de Havilland's body washed ashore.

As an airplane pushes its way through the sky it must necessarily shove aside the molecules of air that occupy the space it is invading. Since sound is actually the motion of molecules, it follows that the fastest speed with which the air molecules can get out of the way is the speed of sound. If an airplane tries to push its way through these molecules faster than they can get away, the inevitable result is that they pile up in a chaotic "wall" in front of the airplane; this is the sound barrier.

Unsuspected at first, it soon came to be regarded as an insuperable barrier. But in just another few years the aeronautical engineers worked out a design feature, based on the German sweptwing Messerschmitt 262 jet fighter, that allowed the piled-up molecules of air to spill back along the wing and be dispersed, causing nothing more than what we now call the "sonic boom," and the way was open to supersonic travel.

By 1966 the United States, Britain, France, and Russia were all planning to inaugurate the era of commercial supersonic transportation. Fleets of several hundred SSTs were envisioned, crisscrossing the world in stratospheric flight.

Stratospheric flight, that was the problem. We mentioned earlier that at the base of the stratosphere is a layer of ozone, which absorbs sunlight and causes an increase in temperature as height increases from that point outward. The ozone actually absorbs not only visible light from the sun, but light of shorter wavelength, that is, light in the ultraviolet region of the spectrum.

Different regions of the electromagnetic spectrum differ in more than our eyes' ability to detect them. X-rays, for example, consist of electromagnetic radiation of extremely short wavelength, and quantum mechanics tells us there is an inverse relation between wavelength and energy. The short-wavelength X-rays carry enough energy to penetrate our skin and soft tissues, making them valuable diagnostic tools but also dangers to our health. Gamma rays, one of the components of radioactive decay, have even shorter wavelengths and thus carry more energy per ray, or more properly per quantum, and are even more penetrating and more dangerous to our health.

The ultraviolet region of the spectrum has wavelengths just a bit shorter than visible light; in Figure 1 it lies just to the left of the radiation labeled "visible." Each quantum of ultraviolet light therefore carries a bit more energy than visible light—not enough to penetrate our skin, but enough to damage that which it touches. The UV region is broken into three subregimes for identification purposes. Ultraviolet-a, or UV-A, just below visible light, has wavelengths of 320 to 400 nanometers; these wavelengths are short enough to render it invisible, but not short enough to make it harmful to us, and so it is of no consequence. Ultraviolet-c, UV-C, with wavelengths of 200 to 280 nanometers, is indeed harmful; in fact, even a small amount of such radiation is lethal to man and other animals, and even to plants. Luckily, it is almost entirely absorbed by the ozone layer. (Actually, it is almost certainly wrong to speak of this as being due to luck. Life on earth probably evolved in the seas, primarily because there was no ozone layer in the beginning, and water can also absorb the UV, which would otherwise have destroyed the first living molecules as quickly as they were formed. It wasn't until sufficient plant life existed to convert carbon dioxide photosynthetically into molecular oxygen—from which ozone is formed—that living creatures were able to crawl out of the seas and inhabit the land.)

Ultraviolet-b, UV-B, between 280 and 320 nanometers, is also lethal to living creatures, and while most of it is absorbed by the ozone, some gets through, causing not only sunburn but also skin cancers and cataracts in our eyes. If the ozone were to disappear, the UV-B and UV-C would get through and all life on earth would probably be destroyed. Total destruction of the ozone is hardly a likely circumstance, but even a small decrease would mean immediate increases in ultraviolet radiation at the surface of the earth, with concomitant increases in skin cancers and cataracts as well as decreased productivity of some of the plants on which we depend for food.

And so the sonic booms of the SSTs were as a corps of buglers, sounding the call to arms for a few people who thought they knew what would follow.

In 1966 Dr. J. Hampson of the Canadian Air Research Defense Establishment published a technical note pointing out that the envisioned fleet of several hundred SSTs would be flying at altitudes within the stratosphere, that is, within the ozone layer, and that their exhausts would include water vapor, a gas normally thought of as innocuous but one which was known to react with ozone. "React with" means that the water vapor combines with the ozone to form something else; and "something else" means that the ozone no longer is ozone. And that means that more ultraviolet would get through to the surface of the earth, where we live.

The Boeing Aircraft Corporation was one of the American companies working on the design of a supersonic transport. They couldn't have wanted to know about any ozone problem, just as the cigarette companies didn't want to know about the cancer problem, but they began a research program to find out what was going on. Halstead Harrison of Boeing's Scientific Research Laboratories initially estimated not only that Hampson was right but that the reduction in ozone would amount to a catastrophic 20 percent, although by the time he published his results in 1970 he had lowered that estimate to 3.8 percent, an amount Boeing thought to be not terribly significant. Other scientists, however, were already arguing that a few percent reduction would lead to thousands of new cases of skin cancer per year.

By March of 1971 enough concerns had reached the public for Congress to feel the need to call a hearing to discuss whether the United States should proceed with its SST program. The major witness arguing the con case was Professor James McDonald, an atmospheric physicist from the University of Arizona, whose interest in SSTs had begun with a study of their effects on climate. He had concluded that there would be no adverse effect, but when he went on to consider the ozone problem he came to a different conclusion. A depletion of even one percent, he thought, would bring on five to ten thousand new cases of skin cancer every year in the United States alone. Although his estimate was based on early data that examined the relationship between ultraviolet radiation and skin cancer largely in animals or in indirect studies of the population, it is remarkably similar to results much more firmly established today. But his impact on the congressmen was not great, particularly when it was revealed to them that he was an enthusiastic supporter of flying-saucer research, and in fact had previously testified to Congress that he thought recent power failures in New York City were related to UFOs.

The Department of Commerce had already set up an SST advisory panel, and in that same month they met in Boulder, Colorado. A reluctant participant was Harold Johnston of the University of California at Berkeley, a much-respected meteorologist and chemist who wasn't really interested in the problem but was talked into attending. Johnston had earlier done work on the chemistry of nitrous oxides (commonly written NO_x as shorthand for the whole group, NO_2, NO_3, N_2O, and so on) in the atmosphere, and during the course of the meeting he became acquainted with some estimates of ozone destruction by these compounds, which are, like water vapor, ejected as exhaust gases from jet engines. He stayed up nearly all one night to calculate the effects more precisely, and the next day announced to the meeting that two years' operations by a fleet of five hundred SSTs would deplete ozone by at least 10 percent and possibly as much as 90 percent, with catastrophic results for the human race.

James McDonald reminded the participants of what he had told Congress: "The whole history of evolution has been a battle with ultraviolet, and so far we've just barely won." Representatives of Boeing and of the aviation industry argued that the ozone link with skin cancer was only a "theory" (the same argument presented about lung cancer by tobacco manufacturers, and by the creationists about evolution) or was even a lie being promulgated by opponents of progress (who might even be communists).

Then, like a bombshell at the meeting, came the news that the House of Representatives had just cut off all funds for American SST development. This was probably done more for economic than environmental reasons, and it is interesting today to ask whether, had the economics been profitable, it would have been worthwhile to build the SST fleet even if it meant causing tens of thousands of skin cancers every year. The answer, if we go by our decisions in similar matters, is yes. For example, last year in Miami the police initiated a new program: They put patrol cars in the parking lots of cocktail lounges and stopped customers as they drove away. Sobriety tests showed that many of the drivers were drunk, and they were arrested for driving under the influence before they could kill anyone. A great program? Surely. What happened? It was canceled within forty-eight hours. The lounge owners complained to the political leaders that it was bad for business, and so it was stopped. Many if not most of the fifty thousand vehicular deaths every year in the United States are alcohol-related, and many could be prevented by such simple measures. But it's bad for business.

At any rate, for economic and perhaps environmental reasons too the SST program was indeed scrapped. The occasional British-French Concorde flights are not enough to cause any discernible effect on the ozone, and so the battle was won. But the war between ultraviolet and humanity was and is far from over.

Refrigeration and air-conditioning are based on two simple laws of physics. One is that heat will always flow from hotter to colder bodies, the other is that heat is absorbed by liquids vaporizing and is released by gases liquefying. For example, if you heat a pot of water, the heat from the stove flowing into the water raises its temperature until it reaches 100° C., at which point the rise in temperature stops even though the stove is kept hot. The boiling water will stay at 100° while the water boils: the heat flowing into the water no longer raises its temperature, but is absorbed in order to vaporize the liquid. When the vapor condenses elsewhere in the room, that heat is returned to the air.

Both refrigeration and air-conditioning are accomplished by getting ambient heat to flow into a liquid, thus vaporizing it. The gas is then transported out of the system and allowed to liquefy, putting the heat back somewhere else. In a refrigerator the heat simply is blown from the inside of the box to the room in which the refrigerator sits; in air-conditioning it is transferred from inside the house to the outside air. In both cases an easily vaporizable fluid is needed.

In the early years of this century ammonia, sulfur dioxide, and methyl chloride were the most-used fluids, but there were problems with each of them. Ammonia and sulfur dioxide are both corrosive, thus likely to develop leaks in the apparatus, and they cause health problems when people are exposed to them; methyl chloride is less corrosive and thus less likely to leak out, but when it does it is not only toxic but flammable, even explosive.

In 1928 Thomas Midgley, an industrial organic chemist at General Motors' research laboratory in Detroit, discovered two inert, nontoxic, nonflammable gases with excellent refrigerant properties: dichlorodifluoromethane and trichloromonofluoromethane. These are simple variants of methane, the proper name for what is commercially called "natural gas." Methane's formula is CH_4, which denotes that the molecule is composed of one carbon atom attached to four hydrogens. In dichlorodifluoromethane the four hydrogens are replaced by two chlorine atoms and two fluorines; its formula is CCl_2F_2. In trichloromonofluoromethane the four hydrogens are replaced by three chlorines and one fluorine: CCl_3F. Together the chemicals have become known as chlorofluorocarbons, CFCs for short. By 1938 the new compounds, marketed by du Pont under the trade name Freon, had captured 15 percent of the refrigerant market, and were rapidly capturing more.

Du Pont investigated their properties and uses even further, and found that they fully justified the du Pont motto *Better things for better living through chemistry.* The compounds could be easily liquefied at room temperature by simply applying pressure, and would vaporize instantaneously with the release of that pressure. They were totally inert; in particular they would not react with water or with human sweat, saliva, skin or hair. Putting these two properties together meant that they were perfect propellants for insecticides.

In the 1930s household insecticides were dispensed with a long tube-and-plunger combination, attached to a receptacle. The poison that killed the insects was a liquid that was vaporizable, but not easily; and so it took a long-handled plunger to reduce the atmospheric pressure in the receptacle, vaporize the poison, and spray it out. Now the du Pont chemists found they were able to mix the poison with Freon and package it in a simple can under pressure, so that the Freon was liquefied. Using a simple dispensing valve invented by Robert Abplanalp (of Nixon fame), a simple push of the finger released enough pressure to vaporize the Freon, which then gushed out in a spray, carrying with it enough droplets of poison to kill the insects. By 1947 forty-five million cans of the new Freon-spray insecticides were being sold each year.

And that was only the beginning. The Freon propellants were totally inert, totally nontoxic. They could be sprayed directly onto people and not hurt them in the slightest. Now what, the du Pont chemists asked themselves, would we want to spray onto people? The answer was obvious. For what is wrong with people? They stink. (The word "smell" is often used as a euphemism, but it is used incorrectly, as exemplified by the story of the disgusted lady who wrinkled her nose and exclaimed to Dr. Samuel Johnson, a wonderful lexicographer but one whose wont was not to bathe regularly, "Sir, you smell!" Dr. Johnson corrected her: "No, Madam, *you* smell. I stink.")

All of us, in an unwashed state, stink. And how many times a day can a modern working person bathe? (In 1914 an unknown Indian mathematician named Ramanujan, who was later to achieve world renown, came to Cambridge to study with the estimable professor G. H. Hardy. At a teaparty the discussion turned to socialism, and one of the ladies present remarked that the problem with the working classes was that they did not bathe often enough, sometimes not even once a week. Seeing the disgust on Ramanujan's face, she quickly assured him that all the Englishmen he was likely to come into contact with bathed every day. With a pitiable attempt at concealing his horror, Ramanujan asked incredulously, "You mean you bathe only *once* a day?" He had wondered what the unaccustomed odor in England was; now, unhappily, he knew.)

The predicament of the great unwashed was a bonanza for du Pont. The miracle of Freon enabled deodorants to be sprayed from pleasantly packaged cans directly onto the human body with no ill effects. There could be no possible environmental problem since the sprays were inert, and if they didn't react with anything how could they harm anything? Environmental safety for once came together with huge profits, and everyone was happy. Soon the Freon sprays were being used not only for deodorants but for hair sprays and a multitude of lesser things. In 1954, 188 million cans were sold in the United States alone. By 1958 the total had climbed to more than 500 million, and by 1968 the industry was selling 2.3 billion cans a year.

Then in 1970 a single British chemist, working alone and alienated from the commercial and university systems, found traces of the CFCs in the atmosphere—and things would never again be quite the same.

Jim Lovelock is best known for his concept of Gaia, which supposes that the earth is in some respects similar to a single living organism, or alternatively that its living organisms control the earthly environment much as any creature controls its own internal environment. This concept is frowned upon by many scientists, is regarded as provocative and therefore useful by others, and is even considered by some as being possibly true.

In 1960 Jim Lovelock was a respected researcher holding a Ph.D. and a tenured position with the British National Institute for Medical Research. In 1970 he was out of a job and working from his private home in the country village of Bowerchalke.

He hadn't been fired; he had quit. He hadn't quit because of any argument with the establishment; he had simply quit. He left a position guaranteed for a lifetime, with a decent pension at the end of it—for precisely those reasons. He quit for the reasons that other people *accept* jobs: security, regularity, a sense of belonging to the group. He took a long look at himself and at his life and decided that those attributes stifle creativity and inhibit fun, and that's what science is supposed to be all about.

So he walked away from the Institute for Medical Research and set up a small lab in his country home to do consulting research. That meant he would get an idea, write a proposal to do the research, and flog it around the country to see if anyone was willing to support it. If they were, they'd pay him to do it. If he didn't get any ideas, or if no one wanted to support them, he and his wife would starve. Not an easy life but better than getting a regular paycheck every month, he thought. (Oh well, it takes all kinds.)

In 1970 Lovelock got one of his ideas: he thought he might be able to measure CFCs in the atmosphere. How? The property of the CFCs that made them so useful to du Pont for such a variety of purposes from refrigerants to deodorants was their chemical inertness: they didn't react with anything and therefore didn't affect anything adversely. This same property meant that they wouldn't be destroyed by any chemical reaction, which meant that the stuff sprayed and leaked into the atmosphere wouldn't be destroyed at all. The CFCs weren't soluble in water, for example, so they couldn't be incorporated into raindrops and rained out of the air or dissolved in the ocean. They wouldn't be oxidized by the oxygen in the air or reduced by the hydrogen, wouldn't be affected by sulfur dioxide or pollutants or ozone or anything else, and so all the CFC molecules that ever had been injected into the atmosphere ought still to be there. He made a quick calculation of how many spray cans had been sold and how many refrigerators had been junked, releasing their internal refrigerant gases, and found that even with all this there would still be just a very minor amount, what scientists call a trace, of the gases in the air. But he had already invented a new device capable of measuring such small amounts of trace gases, and so he thought he'd like to give it a try.

Why? That was a tougher question to answer. Why should anyone want to measure the exact concentration of the CFCs in the atmosphere? The very property that made it likely that they still existed at all, their chemical inertness, made them unimportant and therefore uninteresting.

The real reason he wanted to measure them was simply that he could. He had this great little electron-capture instrument capable of doing the job and no one else could do it, and so he wanted to. Really, that was all there was to it. For Lovelock it was a good enough reason, but he could see that it wouldn't be good enough for anyone else. And so he came up with the idea that since these compounds were manmade and hadn't existed in any abundance before the 1950s, they could be used as tracers for atmospheric circulation.

Tracers are used to study all sorts of processes. In attacking the question of atmospheric circulation patterns, for example, what do you do? You want to know how the air, which is basically oxygen and nitrogen, moves around the planet. But you can't simply go to Tibet, for example, and measure the oxygen there; that wouldn't tell you what you want to know, because the oxygen might have been blown in from Europe or from the Indian Ocean and you'd never know it because the indigenous Tibetan air is also composed largely of oxygen. But if you injected something into the air over Europe—something that wasn't in the air over the Indian Ocean—and that something showed up in the air over Tibet, then you'd know where the Tibetan air was being circulated from. The "something" would be a tracer, and there aren't many compounds that you can inject into the air for these purposes because they'd either react with other gases in the air before they got around the world or they'd be harmful to something and rightfully be banned by the government.

So Lovelock suggested that the CFCs would be useful tracers. For example, since (in 1970) they had been produced in quantity for only the last fifteen years or so, and were used mostly in the northern hemisphere, if they showed up in the air over Antarctica we could compare the concentrations there to those over the Arctic and learn something about how the atmosphere circulates between the two hemispheres.

It sounded interesting, but not interesting enough. No one wanted to fund the research, primarily because the expected atmospheric concentration of the CFCs was so low that it wouldn't be easy to detect them, and Lovelock wasn't able to convince anyone that he could really do the job with his instrument. So after a series of disappointments he sat in the kitchen of his country home in Bowerchalke with his wife, and discussed the problem. Together they decided to go ahead and do it even if no one would pay him.

He bought the necessary components with their household money and set up the equipment in the backyard, and it worked. It worked so well, in fact, that he and his wife had to stop using all spray cans because their emissions flooded his instrument. And so he was able to convince a few people that he had the ability to detect the CFCs, and he got funded to go to sea and measure them far from land to see if they really floated around in the atmosphere for great distances.

In the next year, and again in 1972, he sailed to Antarctica with his instruments and looked for the CFCs all the way there and back.

And he found them. Lots of them.

7

A Planetary Time Bomb

The consequences could disrupt, and perhaps destroy, the biological systems of the earth.

—Paul Brodeur, in *The New Yorker,* 1986

In 1973 Mario J. Molina, now a senior researcher at the Jet Propulsion Laboratory, was a young Mexican who had just finished his Ph.D. work at the University of California at Berkeley. His thesis project had been in the field of laser chemistry, but he accepted a postdoctoral position with Professor F. Sherry Rowland at the Irvine campus of the UC even though Rowland's specialty was the quite different subject of radiochemistry.

As it turned out, Molina was not to learn much radiochemistry. Rowland had reached a point many scientists do in midlife. He was a good, solid researcher, well funded by federal agencies and respected by his colleagues, but not yet in the Nobel class and by this time in his life not likely ever to reach it. I remember meeting him once or twice at nuclear-chemistry conferences in the 1960s, a tall, broad-shouldered, athletic man with hair worn rather long and even then beginning to gray. He was one of the first to carry the new hand calculators around in a belt-slung holster, and was one of the leaders in radiochemistry.

But by the early 1970s he was beginning to look for something else to do, a different line of research to learn, something to stimulate him the way radiochemistry used to do. He planned a sabbatical for the second half of the coming year, to give him a chance to visit another laboratory and learn another subject, so that when Mario Molina came to Irvine to begin his postdoctoral year in October of 1973 Sherry Rowland was mentally packing up to leave.

This is a problem not infrequently encountered by postdocs. When I got my Ph.D. in nuclear chemistry I accepted a postdoctoral position at Brookhaven National Laboratory to work on nuclear reactions under one of the world's leaders, Gerhart Friedlander. But when I showed up a few months later I found he had left for the year to go to Switzerland, and I had to scrounge around to find someone else to work with.

In Molina's case it wasn't quite so bad. Rowland wouldn't be leaving till February, and so the two of them had time to talk about how Molina might spend his year. Rowland suggested that it might be worthwhile to take a look at the CFCs. A few years before, when he was just beginning to look for a new area of research, he had attended an International Atomic Energy Agency meeting in Salzburg, Austria, dealing with the applications of radioactivity to the environment. Nothing in particular clicked at that meeting, but after it he had a conversation with William Marlow of the United States Atomic Energy Commission, the agency that funded his research. During this conversation Rowland mentioned that he was looking for new areas to explore, and so in 1972, when Marlow was organizing a meeting in Fort Lauderdale that would deal with atmospheric problems, he invited Rowland.

Jim Lovelock's measurements of atmospheric CFCs had just been done, but weren't being formally presented at the meeting. Nevertheless, someone mentioned them during one of the informal coffee-break conversations. The scuttlebutt was, in fact, that the concentrations Lovelock had measured seemed to indicate an atmospheric load just about equal to the sum total of all the CFCs so far produced, which meant that the general consensus about them was correct: nothing interfered with them, nothing reacted with them, and once they got into the atmosphere they stayed there forever.

In the fall of 1973, when he was discussing with Molina what the young postdoc could be doing while he was away on sabbatical, Rowland brought up this topic and pointed out that nothing—neither diamonds nor love nor CFCs—are forever. Diamonds in fact slowly but irrevocably transform into graphite on this earth, and we all know about love, but what about the CFCs? *Something* has to happen to them. He suggested that Molina might investigate what their ultimate fate could be, while he (Rowland) was off on his sabbatical looking for something more interesting that they both might do the following year. "Of course," he mentioned, "once they get up into the stratosphere they'll be dissociated by the ultraviolet, if nothing happens to them before."

And so Molina sat down and began to learn about atmospheric chemistry and about the chemistry of chlorofluorocarbons; he began to read books and to search through the more recent journal literature, and he began to compile a list of possible atmospheric sinks.

A "sink" is something that will absorb, eat up, or in some manner dispose of the subject under consideration. In the last chapter we mentioned a few possible sinks for the CFCs that turned out not to work: they wouldn't dissolve in rain water or the oceans, nor would they be oxidized or reduced or attacked by acids. Now Molina found that they also wouldn't be ingested and metabolized by any living creature, nor would they be deposited and buried in sediments. The longer he studied the literature, the longer became the list of sinks that simply weren't; and in the list of sinks that *were* he was able to write down just the one process already suggested by Rowland.

If nothing else happened to the CFCs, he agreed, they would float around in the atmosphere and gradually disperse both horizontally and vertically. Horizontally meant they would spread from the industrialized nations where they were being sprayed and leaked into the atmosphere, and in time they would be all over the globe; as Lovelock's work had shown, this was already happening.

Vertically meant they would migrate upward. Since they are heavy compounds relative to the nitrogen and oxygen molecules that comprise most of the atmosphere, their greatest concentration would remain close to the ground, but they would be driven inevitably higher as their lower concentrations increased, according to simple laws of diffusion. Eventually—if nothing else destroyed them—they would reach the stratosphere and mingle with the ozone there.

Ozone, although a reactive gas, is not reactive enough to react with such an inert molecule. But as the CFCs rose through the ozone they would be exposed to increasing fluxes of ultraviolet radiation—radiation that would be absent below the ozone layer but that was present in increasing doses as the molecules mounted higher and higher. The energy carried by the ultraviolet is enough to dissociate the CFC molecule, breaking it into its components and in particular releasing chlorine atoms. The chlorine atoms are extremely reactive, and so is ozone; obviously the two will react, and the result will be the destruction of the ozone molecule.

When Molina reported this to Rowland, they immediately sat down to calculate the amount of destruction that might be done to the ozone layer if all the CFCs ever produced were to migrate up there, become dissociated by the ultraviolet flux, and produce chlorine, which would then eat up the ozone. The answer was reassuring to civilization, though it must have been a disappointment to Molina and Rowland, who were, after all, trying to find an important research problem. There was nowhere near enough of the CFCs to produce any noticeable effect, nor was there likely to be even if their production should increase, and so this reaction wasn't going to be important.

And somewhere, whatever gods may be must have been laughing.

Well, it happens all the time. The very nature of scientific research means that you're frequently trying to investigate a phenomenon that nobody knows anything about, which means that you can't tell in advance whether the problem you're attacking is solvable or, if it is, whether the solution will turn out to be interesting or not. During my postdoctoral at Brookhaven I became interested in meteoritics, one of the problems of which was the age of iron meteorites. A recent experiment had showed them to be about ten billion years old, but the age of the solar system was well established from stone meteorites at four and a half billion years. It might have been a bit older, when you throw in all possible errors; it might have been 4.6 billion, for example, but couldn't possibly have been any older than 5 billion. Ten billion was out of the question. Of course the iron meteorites could have come from beyond the solar system, which certainly would have been an interesting solution, but this was extremely unlikely on dynamic grounds. On the other hand, it was also extremely unlikely that anything originating within the solar system could possibly be ten billion years old, so it seemed like an interesting problem and I worked on it, off and on, for the next several years. Finally Lou Rancitelli, one of my Ph.D. students, and I discovered that the apparent age of the iron meteorites was due to a totally unexpected phenomenon. Potassium, which as the radioactive progenitor of argon provided half of the age-measuring tool, was leached out of the meteorites during their time on earth, thus resetting the isotopic clock erroneously. The measured age had no real meaning: a totally uninteresting result.

And so Rowland and Molina prepared to write up their investigation of the atmospheric fate of the CFCs along with its totally uninteresting result: photodissociation in the stratosphere, period, end of story. But the gods must have chuckled a bit too loudly, and Molina must have heard them, for instead of writing *finis* to their paper he started to work out just exactly what would happen to the chlorine atoms released by the CFC photodissociation. And by late that night he was hearing the gods loud and clear.

The next morning he discussed with Rowland what he had come up with. Molecules in the stratosphere are exposed to the ultraviolet flux and will dissociate, as they already knew. The CFCs will liberate, among other atoms, free chlorine (Cl); and also, Molina now pointed out, normal oxygen molecules (O_2) will dissociate into free oxygen atoms (O). The chlorine atoms will then react with the ozone (O_3) to form a chlorine-oxygen compound (ClO) plus molecular oxygen. But something else will also happen: the ClO is reactive enough to interact with the free atomic oxygen to form chlorine atoms and molecular oxygen. The sequence can be written in the shorthand notation that chemists use as:

$$Cl + O_3 = ClO + O_2$$

and

$$ClO + O = Cl + O_2$$

It was now a simple matter of adding the equations together: the Cls and the ClOs cancel from opposite sides of the equations and Molina was left with the simple relation:

$$O_3 + O = 2O_2$$

The astounding but inescapable conclusion was that, in the presence of free chlorine and oxygen atoms, the ozone would react with the oxygen to form normal molecular oxygen. The chlorine, although initiating the reaction, doesn't appear in the final equation and so doesn't react at all, in a very important sense. Another way of looking at it is that in the first reaction the atomic chlorine disappears to form ClO, but in the next reaction the ClO gives back the original chlorine.

With either way of looking at it, the point is that the atomic chlorine (Cl) doesn't disappear, isn't chewed up; although it reacts it is also produced, and the net effect is that the single atom of chlorine is still there after the ozone molecule has been eaten up.

This is a well-known phenomenon in other areas of chemistry. It is known as a catalytic process inducing a chain reaction. The chlorine catalyzes the reaction of ozone with atomic oxygen, in which both of them are transformed into molecular oxygen, but the chlorine itself is not affected. This means that each atom of chlorine produced by the photodissociation of a CFC molecule will not only stimulate the destruction of an ozone molecule but will remain in the atmosphere to do the same thing all over again.

In fact it will do it over and over again, until finally the chlorine is eaten up by another reaction and disappears from the scene. Using reaction-rate kinetics, Rowland and Molina were able to calculate that each chlorine atom would aid in the destruction of 100,000 ozone molecules. This meant that their original conclusion was totally wrong. They had thought that there were not enough CFC molecules being produced and sprayed into the atmosphere to eat up any discernible amount of ozone, but now they found that each CFC molecule would destroy 100,000 times as many ozone molecules as they had previously calculated. Instead of being orders of magnitude too few, it now appeared that there were indeed enough CFCs around to affect the ozone layer by at least a few percent.

And suddenly it was a whole new ball game.

The catalytic chain reaction induced by chlorine atoms in the stratosphere had actually been known before Molina and Rowland "discovered" it. Other groups of atmospheric chemists, worrying about the effect of SSTs on the ozone, had known about it but had dismissed it from their calculations because they didn't think there was any source of chlorine that would make it important. Now, of course, there was.

Molina and Rowland wrote up their results and submitted them in a paper to *Nature,* the British scientific journal. (There is no significance to their submitting to a British instead of an American journal. Science is international in this respect, and the two leading weekly journals of interdisciplinary science are the American *Science* and the British *Nature.* They probably sent their work off to *Nature* because *Science* is not really as interdisciplinary as it pretends, publishing almost entirely only biological reports.) They also sent an abstract to the American Chemical Society, indicating their intention of presenting the work orally at the society's Atlantic City meeting the following September. Then Rowland packed up and went off to Vienna for the spring semester.

The paper was published in *Nature* in June, but even before that, word began to circulate, as it often does in scientific circles. Rowland had heard about Lovelock's measurements before publication, in a coffee-break conversation, and in a similar manner word of the ozone problem began to leak around the world, worrying the people at E. I. du Pont de Nemours & Company.

The ozone layer, which shields us from ultraviolet, is not a constant thing, but is continually being produced and destroyed. It is produced naturally by the action of sunlight on normal oxygen molecules in the upper atmosphere, and is destroyed naturally through the action of nitrogen oxides (NO_x), which circulate upward upon being released by bacteria in the soil. The constant production and depletion lead to a steady-state situation in which the ozone is being produced at the same rate as it is being destroyed, and so the level in the atmosphere remains constant—at a concentration large enough to absorb all the UV-C and most of the UV-B. If anything should disturb this fragile equilibrium and deplete the ozone a bit faster, its steady-state concentration would drop.

In their *Nature* paper, Molina and Rowland were careful not to sound a precipitous alarm. They said only that "important consequences may result" and that "more accurate estimates . . . need to be made . . . in order to ascertain the levels of possible onset of environmental problems."

Mild enough. But as word of their work began to circulate through the scientific community, the CFC manufacturers began to feel the first touches of apprehension. They were sailing along on a many-billion-dollars-a-year boat, and they didn't appreciate its being rocked.

So even though the Molina-Rowland paper was mild enough, its implications were worrisome. In fact, there were rumors spreading that Harold Johnston, the Berkeley chemist who had first attacked the NO_x problem in regard to the SSTs, had made some calculations based on the Molina-Rowland work that indicated that the effect on the ozone layer could be a depletion of as much as 20 to 40 percent. And in the April 5 issue of *Science,* Pythagoras Cutchis of the Institute for Defense Analyses in Arlington, Virginia (who was unaware of the Molina-Rowland idea but was still worried about the NO_x effects of a fleet of SSTs), calculated that a depletion at the upper end of that range would lead to nearly a tripling of the UV-B radiation at the surface of the earth. It had been calculated by several other workers that increases of only a few percent would lead to tens of thousands of skin cancers every year. Of course, skin cancer is not the worst kind of cancer; most cases are curable, but still . . . *Better things for better living through chemistry?*

When officials of the American Chemical Society prepicked the Molina-Rowland paper as one of the most newsworthy at the upcoming Atlantic City meeting, guaranteeing press attention, the industry thought it was time to take a hand. They had actually become involved in the problem several years previously. It's all too easy to blame all our problems on the rapacity and greed of big business, and though there's something to be said for the difficulty of overestimating these effects, there is usually something to be said on the other side.

The industry was in fact neither ignorant of nor uninterested in the possible environmental problems. In 1972 du Pont had organized a seminar on the ecological effect of CFCs, stating openly that the world was producing these things at an increasing rate and that it was time they found out if there were any harmful effects "now or in the future." The company had been instrumental in organizing a coalition of CFC producers to sponsor independent research on the problem, and by 1974 the results had indicated no detrimental ecological effects.

So what—they now asked—was all the fuss about?

An industry spokesman called Dorothy Smith, who as news manager for the ACS was arranging a press conference at which reporters could meet with Sherry Rowland, and he put the question to her. After all, he said, there was no proof of the Rowland calculations, no data at all that indicated a threat to the ozone. The whole idea was "just a theory." The phone call worried Smith; the last thing the ACS wanted was to get embroiled with a spectacular story that might later prove to be nonsense. But after calling a few atmospheric scientists, who unanimously assured her that the "theory" was a good one, she went ahead with the press conference.

And, in a manner of speaking, the Flit hit the fan.

At the meeting and press conference Rowland told his audience that if the CFCs continued to grow at their present rate of 10 percent per year until 1990, and then remained constant, there would be a 5 to 7 percent destruction of the ozone by 1995, with an additional and unavoidable destruction of 30 to 50 percent by 2050. (This future destruction is an especially pernicious aspect of the problem. Because the CFCs are so inert, once introduced into the atmosphere they are not removed by any natural process; because they diffuse upward so slowly, it takes years for them to reach the stratosphere. Putting these two facts together, it follows that even if we were to cease all CFC production today, the CFCs already in the atmosphere would slowly but inexorably climb to the stratosphere and there do their damage. This means that even if we see no discernible ozone damage today the damage tomorrow *by the CFCs already in the atmosphere* will be greater and possibly significant. This, in turn, means that if we wait to cut back on production and use until we actually see a detrimental effect, it will be too late: a further drastic increase in the effect is mandatory. What we have here is a planetary time bomb.)

So the effects claimed by Sherry Rowland were not immediately verifiable, since they referred to the future; but they were immediately worrisome. Rowland went on to suggest that even a 5 percent reduction in ozone would cause forty thousand extra cases of skin cancer per year in the United States alone, with similar additional effects all over the world.

In that same month of September 1974, a group at the University of Michigan published computer calculations that supported the Molina-Rowland message, emphasizing that at the present rates the ozone equilibrium in the atmosphere would be displaced downward as early as 1985 or 1990. Another group, at Harvard, published similar work, arguing that "the reduction could be as large as 3 percent by 1980, or 16 percent by 2000, if Freon consumption were to grow at 10 percent per year. Even if Freon use were terminated as early as 1990, it could leave a significant (health, environmental) effect which might endure for several hundred years."

Although the Harvard paper wasn't published until February 1975, it was released to the press the previous September, and on the 26th of that month Walter Sullivan of the *New York Times* wrote a front-page story, and the media and the nation were off and running.

Within a month the National Academy of Sciences had convened an ad hoc panel to make recommendations. Included on the panel were several of the scientists who had rung the bell, including Rowland, Johnston, and Mike McElroy, one of the authors of the Harvard report. Meanwhile, du Pont announced that the industry was policing itself and had begun a comprehensive research program (to be overseen by its own Manufacturing Chemists Association) that would study the problem and all its implications, without fear or favoritism. Of course, doing responsible science does take time, as all men of good will would acknowledge, and the industry wanted to do it properly. Accordingly, they announced, it would take three years to complete the project, during which of course there would be profits as usual. It was to be hoped, they announced, that there would be no panicky reaction (or punitive legislation) before all the results were in.

It was not to be. By the end of that same month of October one of the ad hoc panel was quoted in the *Times* as urging the public to stop buying all spray cans with CFC propellants. The next month the Washington *Post* ran a story saying that the panel would soon be urging the banning of the spray cans. By the end of that month national sales had dropped by 7 percent.

In December the House Subcommittee on Public Health and the Environment opened a hearing, but nothing was done before Congress adjourned eight days later. In January of 1975 James Fletcher, the chief NASA administrator, saw the moving finger writing, and announced that his organization was prepared to act as the government's chief research agency (and pick up the biggest bundle of extra congressional funding) in this problem: "We have rather dramatically changed our emphasis within NASA to focus on stratospheric research."

In January the President's Council on Environmental Quality and the Federal Council for Science and Technology created a task force to conduct an "intensive study" of the situation. The panel included representatives of seven cabinet departments and five different government agencies. In June they issued a report stating that CFCs were a legitimate cause for concern and should be restricted by 1978. At the same time the Senate Committee on Aeronautical and Space Sciences opened their own hearing but accomplished nothing.

The month before the task force met, Congressmen Paul Rogers of Florida and Marvin Esch of Michigan had introduced a bill authorizing the Environmental Protection Agency (EPA) to ban all CFCs if the National Academy of Sciences (NAS) decided the threat was real. Their bill died in committee. In February 1974, they reintroduced it; an even stricter bill was introduced by Les Aspin of Wisconsin. Neither bill passed.

Instead the public heard about a report in the du Pont *Management Bulletin,* which stated proudly that there were no adverse pollutant effects caused by the CFCs in the troposphere. (Well, of course there were none: the whole basis of the ozone problem was that the CFCs didn't react with anything in the troposphere, so how could they cause any problems there? The *Bulletin* didn't discuss the stratospheric problems.)

Still another committee was formed that February: the Ad Hoc Federal Interagency Task Force on the Inadvertent Modification of the Stratosphere (IMOS), sponsored by the President's Council on Environmental Quality, the Federal Council on Environmental Quality, and the Federal Council for Science and Technology. In March the National Academy of Sciences appointed a Panel on Atmospheric Chemistry, to report to its Climatic Impact Committee, which had been established originally to investigate the SST problem (which by now had gone away), and *New Times* quoted Sherry Rowland as saying to his wife when she asked how his day had been: "The work is going well, but it looks like the end of the world."

To which our government's reply was, as we have seen, the formation of more committees. To be fair, it wasn't simple; no one knew who had legal or scientific jurisdiction. Back in November the National Resources Defense Council had petitioned the Consumer Product Safety Commission to outlaw spray cans. There was a law in effect which said that upon such petition the commission must either comply or issue a written denial in four months. They did neither, because they couldn't decide if they had the legal power. The

CPSC general counsel, Michael Brown, suggested, "This may be something for the EPA." But the EPA responded that "We take care of the *lower* atmosphere, and this is an *upper* atmosphere problem." They suggested perhaps NOAA might be the right organization to take charge.

In April Congressman Robert Michel (Illinois) complained that he couldn't understand the whole thing.

Still, progress was made. In June 1975, came the first break in the industry's united front. (They had been reasonable, according to their own lights. They had never rejected the ozone-depletion argument, nor claimed that it was unimportant; they merely claimed that it needed more study. They were investigating it themselves, and in "three years" they could have the answer. This figure, "three years," never changed very much; you have to give them high marks for consistency. In 1974, in 1978, in 1984, they were still sure they would have the answer "in three or two or four years.") But now, in June 1975, the Johnson Wax Company announced it would not use any more CFCs. Since they really hadn't used them in any significant quantities for several years, their decision didn't have much economic impact, but the publicity of their announcement did. Other companies followed, and by that summer the production of aerosol valves had been cut in half.

That June Oregon banned the spray cans altogether, with an effective date of March 1977. New York State decided that by April 1977 all cans would have to carry a warning that the product "may harm the environment." In October the industry fought back, with du Pont taking out full-page newspaper ads saying that their research had turned up new evidence that made the ozone-depletion effect 300 percent less than Rowland and Molina had "estimated." Subsequent work proved that the industry evidence was erroneous; it was simply a case of rushing into print because they *wanted* to believe.

By December, measurements had shown that the CFCs had doubled their atmospheric concentration between 1968 and 1975. The industry now argued once again for a three-year delay for further studies. They claimed that such a delay wouldn't hurt anyone; depletion would be "an insignificant and undetectable effect of one half of one percent." (Even if this was true, and no one at the time knew what the depletion would be, the best estimates were that "one half of one percent" depletion would mean an extra 6,000 cases of cancer per year, and this would continue for roughly a hundred years, so the "insignificant" effect would translate literally into 600,000 extra cases of cancer.)

The public was not fooled. Purchases of spray cans continued to drop, companies continued to desert the market, and by the end of the year Washington was abounding in rumors that legislation banning all nonessential uses of CFCs was just around the corner. The battle seemed to have been won.

The battle, yes; the war, no. For the always fickle attention of the public and the legislators now began to stray toward a new romance.

8

The End of the World: Part 2

The inhabitants of planet earth are quietly conducting a gigantic environmental experiment. So vast and so sweeping will be the impacts of this experiment that, were it brought before any responsible council for approval, it would be firmly rejected as having potentially dangerous consequences. Yet, the experiment goes on. . . .

—Wally Broecker, geochemist at Lamont-Doherty Geological Observatory

On September 27, 1974, *Science* carried a news article about researchers' efforts to perfect computer models describing the earth's climate and thus to predict future weather. After pointing out the recent famine in drought-stricken regions of Africa and concern over the size of the current U.S. grain crop, they concluded that the world's food supply was now perilously dependent on the unlikely prospect that the global climate would continue in the same general pattern seen for the last hundred years. "Abrupt changes in the length of the growing season, annual rainfall, or mean temperature could disrupt modern and primitive agricultural systems alike," they reported. "Unfortunately," they continued, "shifts in the earth's climate appear to be the rule rather than the exception."

The article went on to warn that though the earth's climate has recently been unusually warm and friendly, this trend was unlikely to continue, according to recent scientific results. Instead, a trend toward cooler temperatures had been discerned, they reported, and warned that "a significant global shift in climate may be under way—a shift that could be the forerunner of a new ice age."

The article did not mention the ozone problem, although this was two weeks after the American Chemical Society meeting in Atlantic City, and the same issue carried the University of Michigan computer calculations. The article also did not mention the greenhouse effect. It would be the last scientific article about future climate problems not to mention either of these. Although it was correct in predicting a significant global shift in climate, it had the sign of the change totally wrong, thus tying for first place in the Honor Roll of All-Time Worst Predictions with two other classics: William Jennings Bryan, on the occasion of the hundredth-anniversary celebration of the ending of the War of 1812, claimed in early 1914 that "We know of no cause today that cannot better be settled by reason than by war. I think we have seen the last great war," and the president of the New York Stock Exchange said in September 1929, "We are apparently finished and done with economic cycles as we have known them. I see nothing but continued steady growth ahead."

Indeed, our view of the future is a cloudy one, and cycles in climate as well as in economics are difficult to predict and terrible in their consequences.

In 1956, in a paper published in the scientific journal *Tellus,* Gilbert Plass of Johns Hopkins University warned that carbon dioxide controls our climate, and further that it is an unbalanced control lacking in stability: "There is no possible stable state for the climate." He went on to point out that mankind's increasing use of fossil fuels must have increased the atmospheric concentration of carbon dioxide, and that an increased greenhouse warming of the climate must surely follow.

He was not the first to issue this warning. In 1896 the Swedish chemist Svante August Arrhenius, who practically single-handed founded the modern science of physical chemistry, took up the point first noted by the French mathematician Jean Baptiste Fourier—that the carbon dioxide in the atmosphere acted much as did the glass panes of a greenhouse—and warned that if we continued burning our carbon-based fossil fuels so prodigally we might actually increase the global temperature beyond its normal range. At the time, the idea that man might have a noticeable effect on the planet seemed ridiculous; the earth is, after all, so big and we are so small. And though the poet George Herbert had warned three hundred years ago of the strong connection between the nail of a horseshoe and a king's crown, the same people who read Herbert did not read Arrhenius.

In 1938 the English meteorologist G. S. Callendar published a paper in the *Quarterly Notices* of the Royal Meteorological Society, presenting some historical records of atmospheric carbon dioxide and showing that there was a trend of increasing concentrations.

This didn't attract much notice, in the normal way of most research papers; they don't make the newspapers, but a few scientists here and there read them, and possibly remember them, and every once in a while are stimulated by them.

Twenty years later, after a few people around the world had been talking about Callendar's data, Charles David Keeling started a program of measuring carbon dioxide in the air on top of the dormant Hawaiian volcano Mauna Loa, and John Kelley, Jr., was doing the same thing as he puttered along the Arctic Ocean shores of Alaska in an old fishing boat. Far from any industrial pollution, they were able to show that the carbon dioxide concentration was indeed increasing year by year. Before the Industrial Revolution had gotten under way at the end of the nineteenth century there were about 275 parts per million (ppm*) of carbon dioxide in the air; by the time Keeling and Kelley made their measurements it had risen to 311 ppm.

It took another decade for anyone to put the new data together with Arrhenius's warning. In 1967 Syukuro Manabe and Richard Wetherald at the Department of Commerce's Geophysical Fluid Dynamics Laboratory in Princeton, New Jersey, calculated that global temperatures could increase about 2.5° C. if the carbon dioxide concentration doubled. And at the current rate of increase that could happen as early as the beginning of the next century.

And what would a two-and-a-half-degree increase in average global temperature mean? It didn't appear terribly cataclysmic to the average newspaper reader, and so the calculations never made it into the public press, but a few scientists began to take it seriously because of the positive-feedback mechanisms built into our climatic cycles. Would a warming of this magnitude begin to melt the northern and southern ice caps? If that happened, wouldn't less sunlight be reflected away from the earth, causing even more warming? What would be the effect of such warming on weather as it related to agriculture? Would it mean less rain or more? Longer growing seasons, or perhaps a longer breeding season for pests like locusts? What would happen to sea level and our coastlines if the glaciers began to melt world-wide?

There was no panic in the 1960s, but there began to be a growing

*Meaning that if you took one liter of air you would find 275 millionths of a liter (or 0.275 cubic centimeters) of carbon dioxide mixed in with it.

concern, and in 1968 a symposium on the Global Effects of Environmental Pollution was brought together at Dallas. Its primary emphasis was on the carbon dioxide problem. F. S. Johnson expressed the consensus: "Though detailed studies are lacking, the possibility exists that world climate may be affected [by manmade CO_2]. The risk of a serious perturbation appears small, but the problem is only poorly understood and the confidence level in such a prediction is low."

In other words, we didn't know what was going on, but we had damned well better find out. S. Manabe reported numerical calculations that indicated a 0.8° C. increase in temperature by the end of the century. But Reid Bryson at the University of Wisconsin disagreed, arguing that the effect of atmospheric turbidity—the dustiness of the air caused by pollution—"appears to have exceeded the effect of rising carbon dioxide." This meant that the earth would be cooling off rather than warming up; he was still worried about a new ice age. In fact he was warning that man-made pollution was likely to keep increasing and have an even greater cooling effect in the future. Other workers argued that the carbon dioxide was the dominant man-made effect, and that there would be a warming trend for at least the rest of the century.

Well, what did the data show? Since the Industrial Revolution, when the carbon dioxide began to climb and simultaneously the dustiness in the atmosphere increased, what had the global temperature been doing? It's a simple question, and one has the right to expect a simple answer: has the world been warming up or cooling down?

Unfortunately, one's rights in this world are not guaranteed by any sort of celestial constitution, the covenant of the ancient Israelis with God notwithstanding, and the answer to that simple question is decidedly not simple. First, the data: Figure 3.

This shows the carbon dioxide increase since 1760, and the temperature since the first accurate measurements in 1875. (The CO_2 data are in ppm; the temperatures are in arbitrary units relative to 1875, with a total spread of less than 1° C.) From 1875 until 1940 there is an excellent correlation: a rise in temperature as the carbon dioxide increased. But in 1940 the earth definitely began to cool down despite a continuing increase in carbon dioxide; for the past twenty years it has been warming up again. (Actually, *definitely* is a word that should never be used in conjunction with our earth. For

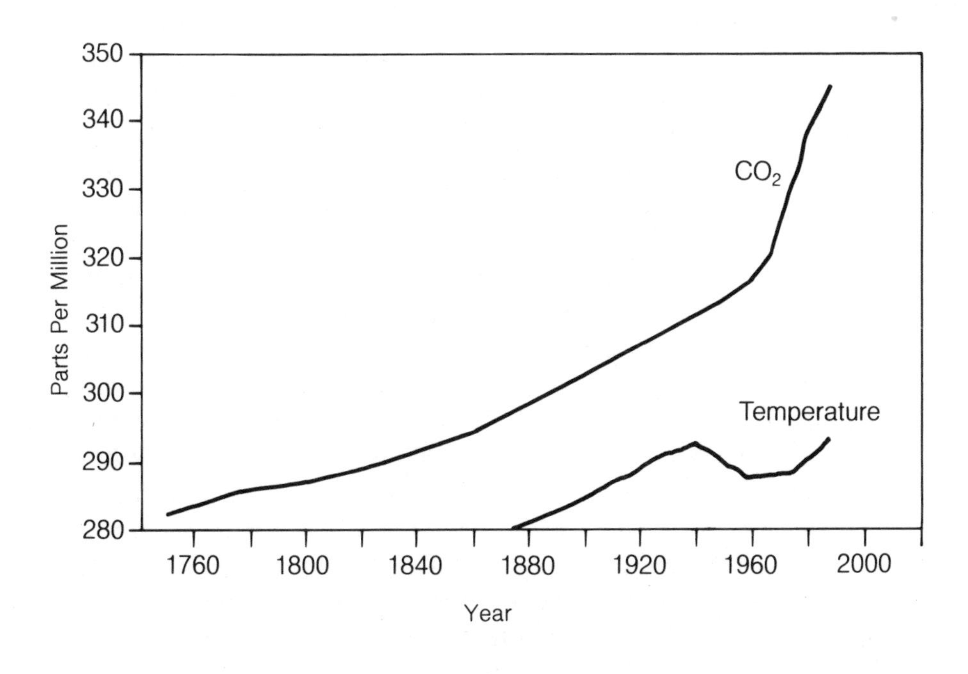

Figure 3

example, the data in this figure come almost entirely from the Northern Hemisphere, for the simple fact is that is where most of the science in this world is done and where most of the industrialization is, and therefore where most of the historical records come from. We might expect atmospheric circulation, at least on a long-time scale, to distribute the carbon dioxide evenly; thus we should be able to discuss the hemispheres separately, with each of them representative of the earth as a whole, but who knows?)

If we take the data at face value, how do we account for the fact that the earth cooled down in the forties and fifties as the amounts of carbon dioxide continued to rise? The first and most obvious interpretation is that as we began to pump the debris from our fossil fuels into the atmosphere, the greenhouse effect was dominant and the temperature began to rise; later, as more smoke, soot, and dust from fuel burning were thrown into the air, they overwhelmed the carbon dioxide, turned the air opaque, deflected sunlight, and began a cooling trend.

Remember, either of these effects has the potential for disaster:

warming could destroy our agriculture and drown our coastlines and even a significant part of our continents, while cooling could bring on another ice age. But before we try to get the situation under control, we have to understand what the situation *is.* What is going on?

Basically, the problem is that the data are not telling us a story as simple as it appears to be on the surface; it's like watching a Pinter play, where all sorts of things are happening that are only dimly revealed. In order to see them more clearly we have to go back in time to before the Industrial Revolution, when the carbon dioxide wasn't increasing and complicating matters.

The interior of Greenland is covered with an icy glacier that has not melted for thousands of years. Just as the sediments on the bottoms of the oceans continually build up from the infall of dead organisms, so the glacier has been built up by the yearly snowfalls. Every winter it snows on that glacier, and year by year the snow is buried by the next year's fall and is compressed into ice by the weight of the years. The deep-sea sediments generally accumulate at the rate of a few centimeters every thousand years, while the glacier accumulates much faster, at the rate of a few centimeters every year. This means that we can decipher shorter time intervals in the Greenland ice than in the oceanic sediments; we can even read off the record year by year because instead of accumulating continually, as the oceanic sediments do, the ice is deposited periodically: every winter there is a new deposit, every summer the process stops. And so the ice is laid down in distinct layers, similar to tree rings, and by drilling into the ice at Camp Century, Greenland, counting off the layers as we go and measuring the isotopic composition of the frozen water and bubbles of carbon dioxide trapped there, we can measure the carbon dioxide composition and temperatures year by year over the past fourteen hundred years.

Two major facts emerge. First, the carbon dioxide content of the atmosphere before the Industrial Revolution was quite constant at about 275 ppm, as we expect. Second, the temperature was not at all constant, contrary to our expectations but not surprising when we look back at our written history. The data show a series of fluctuating temperatures, with a complicated series of cycles of colder and warmer spells, and this is reflected in what we know from the historical record. When I visited friends in Sicily a few years ago, for example, they took me to see a rather unusual sight. At Palermo there stands a bridge, the Ponte dell'Ammitaglio, built in the twelfth century. Such ancient structures are not that unusual in Italy, but this bridge is a magnificent one, spanning five full arches, and the water over which it rises is nothing but a hesitant and tiny stream, worthy of no more than a simple wooden crossing. Why such a giant bridge?

Because in the twelfth century the European climate was much warmer and wetter than it is today. The rivers in Sicily ran faster and deeper, and that small stream was a surging river. Such climatic variations have occurred and recurred and have strongly affected the development of our civilizations.

The first Viking attempt to settle lands across the Atlantic Ocean was in 865, when a Norseman sailed to Iceland; he gave up and returned home, naming the land after the ice he had found covering everything there. But a decade later the climate had warmed sufficiently to allow a successful settlement, and a hundred years later when Erik the Red sailed off to the west he found an island ringed with greenery, which he named Greenland. As a result of his enthusiastic descriptions, a group of Norse families sailed there and tried to form a permanent settlement, but the next few centuries saw a return of colder weather and the death of all the Greenland settlers as the snows closed out the greenery.

In North America the Mill Creek Indian nation had established itself in the prairies and valleys east of the Rocky Mountains in the twelfth and thirteenth centuries, but then the cold came. The weather turned dry, the grassy plains died and could no longer support the buffalo and deer that the Indians hunted; even the forests disappeared. As the crops failed and the animals fled, a stable Indian civilization withered and faltered, and the people reverted to the nomadic life they had led before, following the weather and the animals wherever they led.

During the next few centuries there was drought in Europe too, as more and more water was locked up in northern ice. The Great Fire of London in 1666 was undoubtedly helped by the dryness of the timbers. The Thames would freeze solid from shore to shore in those days, allowing ice-skating parties. The great waves of bubonic plague that swept through Europe seem to have followed upon massive crop failures, undoubtedly due to the terrible weather, which left the population weak and susceptible.

The weather since then has been a series of warming and cooling events, with time scales of from centuries to decades, and they cannot be blamed on the greenhouse effect since the ice cores show that the carbon dioxide composition of the atmosphere did not vary. The reasons are not well understood; they almost certainly are related to variations in the jet streams that circle around the northern latitudes, but the reasons for these variations are not clear, beyond the certainty that they are "natural" variations; that is, they are not caused by our own activities.

Aside from their effect on living conditions as they occur, these natural cycles make it difficult to interpret what is happening today. One attempt was made by Wally Broecker of Lamont-Doherty Geological Observatory in 1975, using cycles that were recognized in the ice-core isotopic data. His model accounted for the rise in temperatures from about 1850 to 1940 *without* any recourse to the greenhouse effect and the increasing carbon dioxide concentrations, and also qualitatively agreed with both the cooling trend that began in the 1940s and the recent rise in temperatures since then.

But the *extent* of the cooling and subsequent warming does not quite fit the expected trend. The earth didn't cool off quite as much as it should have during the cooling period (according to predictions based solely on the cycles), and now looks to be warming up faster than it should. This might be the greenhouse effect we are seeing—or it might be just a failure of the model to predict future weather based on past records. To get a better idea of where the fault lies, we have to take a closer look at atmospheric models and modelers.

9

The Ghost of Climate Future

The farther backward you can look, the farther forward you can see.
—Winston Churchill

Winston Churchill was a master with words, but in the words quoted above he wasn't telling the whole truth. Not only do we have to see back into the past, we also have to understand what we see.

The past climatic record is reasonably clear, but its meaning is not. If the data showed, for example, that the global average yearly temperature was a constant 15° C. for the past thousand years, then today we could say that the 0.5° rise we've seen this century is meaningful and portends even higher temperatures through an effect operating today that wasn't there yesterday, and the greenhouse effect would be the obvious choice. Or if the temperature in the past cycled without complication, if it rose to 16° in the years 1500, 1600, 1700, 1800 and 1900, and dropped to 14° in the midcentury years 1550, 1650, and so on, then today we could say that the warming trend in the latter part of this century is a natural effect, not due to the greenhouse effect, and we could expect that it will warm to 16° and then start cooling again. There wouldn't be much to worry about.

But that isn't the case. The past record is a messy one, with peaks and valleys that show no simple cycle; they may be due to a superposition of cycles, but then again they might be either random or due to cycles we haven't yet pulled out of the data. It's always worrisome to look at a jumbled record and try to see regularities there; sometimes the regularities are figments of our overactive imaginations. This is particularly true when we don't understand the reasons for the cycles we think we see, and so the obvious next step is to try to understand the various causes of climatic variations.

The way to do this is to set up atmospheric models.

Different scientists will give you different answers if you ask them what the basis of science is. Some people say it's the concept of testability: you shouldn't believe anything until you test it, and if an idea is incapable of being tested (such as the reality of God) it isn't a scientifically worthwhile idea (although of course it may be worthwhile from different points of view). Other people focus on the alternation of theory and experiment, which is really just another way of saying the same thing: you get an idea (such as that the sun goes around the earth in a circle) and then you do an experiment (measure the positions of the planets night by night) that tells you that the theory is wrong, so you try a different theory (such as that the planets go around the sun in a circle) and do another experiment that tells you you're wrong, so you try another theory (the planets move in ellipses) that puts you finally on the right track, and eventually you end up with Einstein's theory of relativity—but still you don't believe even that, and you go right on experimenting.

Both these descriptions of the scientific method are valid, but in practical terms they miss the basic point, which is: isolate and simplify. The way to test something, the way to do an experiment, the way to understand what's going on, is to isolate the phenomenon you're interested in and simplify it. This is necessary because the universe is so complicated, with different causes having different effects all over the place, that unless you can artificially separate them you're going to get nothing but confused.

For example, suppose you thought that maybe cigarettes cause lung cancer and pregnancy. You might try to test this by observing who it is that gets lung cancer and pregnant. A generation ago you'd have found that practically everyone with cancer *and* practically everyone with a baby had smoked a cigarette at least once in their lives, and that most of the people without both a baby and lung cancer had never smoked. But if you looked more closely you'd see that the people in the second group were practically all children, so maybe cancer and pregnancies were caused by cigarettes or maybe they were just part of growing up.

In order to answer the question you have to simplify the situation. You have to find people who live all their lives without cigarettes, people who smoke regularly, and people who smoke heavily, and compare these populations. When you do, *then* you find that lung cancer does go along with heavy smoking while pregnancy doesn't, and you're beginning to understand the processes involved. Lots of people make the mistake of not separating and simplifying complexities; they argue that since all heroin addicts first smoked marijuana, and since all perverts have looked at pornography, it follows that marijuana causes heroin addiction and pornography causes perversion. They might as well blame milk for these effects, since both addicts and perverts also drank milk before their drug and sexual addictions began.

If you want to understand the natural variations in climate, you have to isolate the phenomena and simplify the situation. You can't just go out and look at what's been happening, because it's too complicated to understand. There are too many causes causing too many effects, and they interfere with each other, reinforce or cancel each other, hide each other, and generally mess things up. The thing to do, if you could, would be to set up a simplified world: a world without clouds or oceans, for example. Then you could vary the carbon dioxide content and see just how fast the temperature climbs without worrying about things like increased evaporation as the air warms up, which might lead to increased rain or to increased clouds, which might block more sun and so cool things down a bit. Then, once you understood the greenhouse effect perfectly, you could introduce an ocean and see if the increased carbon dioxide would all go into the atmosphere or whether some would dissolve in the oceans, thus lowering the greenhouse effect. And then you might introduce clouds, and little by little you would progress from a simplified world that had little relation with the real world, but which you could understand, to a more complicated world with more and more connection to our real world. That's the way to do an ideal experiment.

If you're a simple person, like a chemist, and you have a simple problem like the temperature dependence of the solubility of silver chloride in pure water, you can do experiments like that. But if you're trying to understand a complex system like the whole earth, you just can't.

So you do the next best thing: you use a computer, and you set up models. You start with the simplest model, something you can understand but which doesn't have much relation to reality, and step by step you introduce complications that move it closer to reality while—hopefully—maintaining your understanding of the situation.

Atmospheric modelers use the box system. For example, a "one box" model would look like Figure 4:

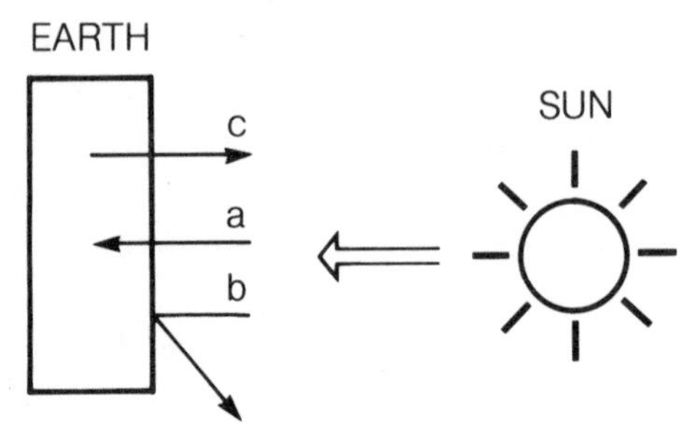

Figure 4

In this model the earth is represented by a single homogeneous box, without considering it to have oceans, an atmosphere, and so on. (The fact that it's represented as a box instead of a sphere is immaterial; that's just the way physicists talk.) The arrows indicate solar energy absorbed by the earth (a), reflected (b), and the earth's heat energy emitted to space (c). With this model you might investigate the effect of changing the sun's temperature; this is the sort of reasoning that led to Figure 1 (the radiation of the sun and earth as functions of temperature). The simplicity of the model makes possible the calculation of the earth's radiation as a function of temperature, but it also means that the answer is going to be only roughly true because we're ignoring so many variables. As it turns out, this model is useful for answering at least one particular question—At what wavelengths will the earth reirradiate its absorbed solar energy?—because the things we've ignored aren't too important in this regard, although they're very important in relation to other questions. This model, for example, would predict that if the sun's temperature rises the temperature of the earth will also rise, and we know that's not necessarily true. When the earth was young the sun was about 30 percent cooler, but as it heated up the earth's temperature remained the same. This was because of a growing greenhouse effect and the various negative feedbacks incorporated into it.

So to understand more about the earth we introduce a more complicated model: a two-box model (Figure 5).

In this model the earth is represented as two boxes, land and air/water. This is closer to reality, because now we can introduce energy transfer between the continents and the ocean/atmosphere, or the different albedos of continents and oceans, and we can see how these factors vary with increasing solar radiation. And we can do something else: we can hold the sun's temperature constant but change the relative proportions of the two boxes—make an earth with more or less land and water—and see how that affects the global temperature. This is getting closer to reality because not only does the real earth have both continents and oceans, but the proportions of each have changed over geological time as the continents have grown and the oceans have spread or receded.

A three-box model might introduce the atmosphere, or perhaps snow, to the earth. And little by little, as our understanding grows, we introduce more boxes and approximate closer and closer to the real world.

The same technique can be considered in regard to the *dimensionality* of models. A zero-dimensional model would be the simplest: this would study what happens at a single point in the atmosphere as various things change. Such a model obviously cannot take account of winds, or motion of any kind in the air, and so is limited in its usefulness. It can, however, tell us something about the effect of different reaction rates—for example as a function of temperature—on the chemical state of the atmosphere.

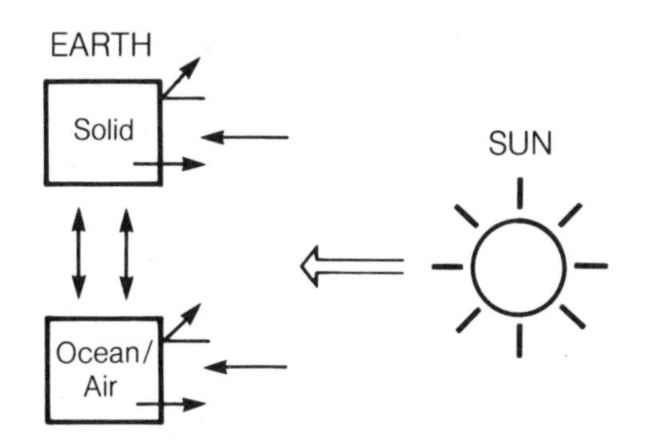

Figure 5

One-dimensional models are an obvious improvement, and are useful for learning about the changes that can occur in a vertical column of air; the effects of rising hot air or sinking cool air can be incorporated here, but longitudinal effects, such as winds, are excluded.

Two-dimensional models can include not only wind effects but changes due to solar insolation as a function of latitude; these models begin to approach reality—and begin to approach the limits of our computer technologies. They can include both physical and chemical dynamics and the effects of the changing seasons, and with them we begin to approach the tradeoff between simplicity and reality. Taken to their most complex, they are reasonable approximations to the real world but are too complicated to evaluate completely within the finite lifetimes we and our computers have to operate in. This is a very real consideration; it doesn't do any good to have a computer model that can solve all the vagaries of weather if it is going to take a hundred million years of computer time to do it. We think of computers as being extraordinarily fast in their computations, and indeed they are compared to our own mental facilities, but these climate models are awesome in their complexity, even to a computer.

Finally we can move to three-dimensional models, which ultimately will reproduce the real world in the electronic insides of a computer, and will tell us everything we want to know—ultimately, but not today or even tomorrow. These models are so complex that only the simplest of them have yet been run, and even these in simplified postures. There are three or four groups around the world that have put together 3-D models, and each of these makes different simplifying assumptions in order to get an answer out of the computer within a reasonable time. So it is no wonder that these various computer models do not give the same answers to the questions we want to ask them.

A typical three-dimensional model requires more than 500 billion computations to simulate the world's climate over one year. The earliest 3-D models, put together in the 1960s, were two-box models. They stipulated a world consisting of one continent and one ocean (the two boxes), with a fixed cloud cover over each, and with no change of seasons. It was a good beginning, but it wasn't very useful. The main thing you want to do with a two-box model is study the interactions between the boxes, but the first models had a major problem: the continents and the oceans both vary with time, but they operate on different time scales, and so the computer couldn't follow their interactions with any reasonable certainty.

By today we've made a lot of progress, and typical models use an enormous number of boxes. The atmosphere may be broken into as many as nine vertical layers, each defining the height of a box, with horizontal dimensions on the order of several hundred miles. To cover an earth whose surface area is about fifty million square miles takes roughly thirteen hundred of these boxes for a single layer, and with nine vertical layers that comes to a more-than-ten-thousand-box model. Even this model, however, isn't much more than a gross approximation to our real world. Each box of several hundred miles on a side, for example, is treated as a homogeneous entity with no internal variations allowed. That's the meaning of "box" in the modeling game, after all. But such a box corresponds to an area on the earth's surface of roughly state-size proportions, and if you live in Massachusetts or Arizona or Florida you know that the weather on any one day is not likely to be the same everywhere in the state.

At the present time these models are being tested by asking them to simulate current conditions, which sounds simple but which takes many man-years and a lot of expensive computer time to accomplish. When a group think they have a model that accounts for our present climate reasonably closely, with what they consider to be reasonable assumptions, they can then increase the carbon dioxide content of the atmosphere and let the computer run until the model reaches equilibrium. The result will tell us what the climate might look like at the postulated increased carbon dioxide levels.

And then again it might not. Until 1988 all the models in operation had to add their increased carbon dioxide in one chunk, and this is another example of where the models take leave of reality. In the real world the carbon dioxide is increasing slowly, and all systems interact as it does, so that as the carbon dioxide begins to increase the temperature rises, which in turn causes more water to evaporate and more surface rocks to be weathered, which will in time cause more ocean sediments to be dissolved, which will in turn affect the amount of carbon dioxide absorbed by the oceans (which of course decreases the effective carbon dioxide in the atmosphere). Meanwhile, more clouds may be forming and thus reflecting away more sunlight; clearly the whole system has a very complex response to a simple data input. This complex response means the world becomes a different place from what it was before the input occurred and so it will respond differently to a further input, so when more carbon dioxide is added it is coming into this new world where new responses are to be expected. But if the model can accept only one-chunk increases, it can't reflect this complex stepwise reaction; the final result may have no relation to reality.

In 1983 James Hansen and his colleagues at the NASA/Goddard Space Flight Center worked out a model that could accept a stepwise addition of carbon dioxide, and could model the earth's reactions as the concentration built up. It took Hansen's computer from 1983 until 1988 to complete the calculations, which tells you something about how complicated the system gets. His is one of the best models now available, but it too has limitations; other models do some things better, but no model does everything better. For example, none of them satisfactorily account for the effect of increased cloud formation as temperatures change. Stephen Schneider of the National Center for Atmospheric Research recently said: "When I first started looking at this in 1972, we didn't know much about the feedback from clouds. We don't know any more now than we did then."

In fact, it's hard even to know when you *don't* know something. Suppose, for example, you're interested in some specific geo-thing, like river flow. Suppose you have a model that says there is a natural "down" direction from the North Pole to the South, such that all rivers flow north to south. You go out to collect some data to test the model, and since you live in North America all the rivers you look at are in this part of the world; you see the Mississippi and the Missouri and the Ohio and the St. Laurence, and they all flow north to south. So you give your paper at a big meeting and everyone congratulates you; your model fits the data well.

But then somebody discovers the Nile. So you modify your model: On the western side of the Atlantic all rivers flow north to south, on the eastern side they flow south to north. In fact, when you think about it, this balances the water flow. It sounds reasonable. You think you've really got a good theory here.

And then somebody discovers the Amazon, and then the Congo and the Niger and the Seine, and you realize your model is just no good at all. Some rivers flow north-south, some south-north, some east-west and some west-east, and there doesn't seem to be any theoretical justification for any of it. And then finally you or somebody in your group begins to understand the true nature of gravity, and you realize that the flow of rivers has no relation to geography but is determined simply by height: all rivers flow downhill, and that's all there is to it.

So your old model was 100 percent completely wrong, because of a misunderstanding of one of the basic concepts of nature. *But* when you first tested it, it agreed with all the data. It wasn't until more data were gathered that you realized it was wrong. This illustrates a basic attribute of science: eventually enough data will be gathered to tell you if your theory is on the right or wrong track. This is the great strength of science. But at any given point in time, just because you have a theory that agrees with all the known data doesn't necessarily mean it's a correct theory. There may be more data hiding somewhere that will prove the theory totally wrong. This is the great weakness of science: you never know for sure.

The best way of testing a theory or a model is not merely to ask if it fits all the data, but to ask what it predicts about the future. Then you wait for the future and see if it's true. In this manner, Einstein's general theory of relativity predicted that light passing close to the sun would be bent from its normal straight-line path, and when the next eclipse occurred a couple of groups observed starlight passing by the sun and saw that indeed it was bent. This bending of starlight constituted the first great proof of general relativity.

But for our climate models the future is too late. If one model predicts a catastrophic warming effect in fifty years and another one says not to worry, we can't afford to wait the fifty years to test them, because if the first one is right our civilization will collapse. We want to know *now* if any of the models are correct.

But all we can do is test them against past data and see how they fit; if they don't fit well, we see their weakness; if they do fit well, we still can't be sure they're any good, for some future data may contradict the model and reveal the world to be either more complex or simply different than we think it is: overwhelming considerations may have been incorrectly estimated, ignored, or not understood at all.

For example, at a conference on the Global Carbon Cycle in 1983 William Jenkins of Woods Hole Oceanographic Institute reminded the audience of some unexpected data he and co-workers from the other major oceanographic institutions (Miami, Lamont, and Scripps) had gathered during cruises of the North Atlantic in 1981. The results, when compared to similar data obtained on a 1972 cruise, showed that during the 1972–81 time period the Atlantic Ocean had undergone large changes in temperature and salinity, which were not predicted by any oceanic model. "If the deep sea can change with such surprising swiftness, presumably in response to atmospheric changes," Jenkins asked, "how will the deep ocean and its capacity to remove carbon dioxide respond to the greenhouse effect?" The answer is clear: we don't know.

One thing that might give us some confidence in the models would be if they all agreed. Since they all make different assumptions and different simplifications, if they all came out with the same conclusions, that would be reasonably good proof that the assumptions and simplifications were valid, or at least unimportant. But in this wicked world such agreement does not often occur, indicating the opposite conclusion: that the assumptions and simplifications *are* important and incorrectly accounted for.

The scientists involved understand this, and put different weight on the conclusions of different models, depending on what aspect of climate modeling they are investigating. Some models handle clouds better than other models, some handle the oceans better, some handle atmospheric currents and some handle chemical dynamics. You pays your money and you takes your choice, and where the models disagree you just don't know what's going on; but perhaps in those respects where they all do agree we might feel some confidence in the results.

So where *do* they all agree?

10

The Long, Hot Summer

There's nothing you can't prove if your outlook is only sufficiently limited.

—Lord Peter Wimsey, in Dorothy Sayers's *Whose Body?*

In 1974 Will Kellogg, who was then associate director of the National Center for Atmospheric Research (NCAR) in Boulder, Colorado, and Stephen Schneider, one of the staff there, published a paper that emphasized the uncertainties introduced by a consideration of clouds in their model. Clouds, they said, are the dominant reflectors of solar energy from the earth. If man supplies heat to the environment, either through increased carbon dioxide and the greenhouse effect or through direct injection by the burning of more fuels, the immediate effect obviously will be that the air will get warmer; but the warmer air might cause more water to be evaporated from the oceans, which would cause more clouds, which would reflect more solar heat, perhaps cooling off the earth. On the other hand, they admitted, the increased water content of the atmosphere might stay in the form of vapor instead of condensing into clouds, in which case the effect would be to increase the greenhouse effect (since water vapor is an efficient greenhouse gas), thus making the earth even warmer. All in all, they reported, this is a very complex world we live in.

On the *other* hand, we all know that salty water doesn't freeze as easily as fresh water. (The dissolved salt lowers the freezing/melting point; that is why we throw salt on icy roads to melt the ice.) If the earth gets warm enough to start the glaciers melting, the water they would send into the oceans would be fresh, not salty (because they form from snowfalls, which in turn form from evaporated water, and salts do not evaporate from the oceans as the water does). This melted glacier water would obviously decrease the overall salinity of the oceans, thus causing *more* ice to form and reversing the effect of the melting glaciers.

As Kellogg and Schneider said, it's a very complex world. At the time, they were more worried about the coming of a new ice age than about the greenhouse effect, and warned that this world came with no guarantees, not even that the sun would stay constant. A "negative change in energy input (from the sun)," they reported, "of the order of 1% . . . could plunge model climates into an ice age."

Discussing the effect of pollution, they pointed out that it too is more complicated than had been thought. Small aerosol particles* can absorb some fraction of the sunlight and scatter away the rest; as they float over the continents, their effect will be to warm the earth since they absorb more light than does the land. But water absorbs more sunlight than either land or the floating particles do, so pollution over the water will cool the earth. "Several studies have been made of the overall effect of aerosols on the radiation or heat balance of the earth-atmosphere system, and given the (meager) available data the consensus appears to be slightly in favor of a cooling," they concluded, while emphasizing that the model results were preliminary and not to be taken as gospel.

By 1977 people were beginning to worry more about the greenhouse effect and global warming than about the next ice age. Two papers in *Science* that year, in April and September, pointed out that we didn't really know all that much about the total carbon dioxide budget of the earth, that is, how much carbon dioxide is generated by fuel burning, how much of that goes into the atmosphere and how much into the oceans, and how much is soaked up by living plants. For the first time, in fact, people were beginning to wonder if the world's forests weren't part of the problem instead of part of the solution.

Plants in general and trees in particular incorporate carbon dioxide from the atmosphere into their bodies, using the carbon and exhaling the oxygen; and so they are a sink for atmospheric carbon dioxide and a source for oxygen. But when they die they are eventually attacked by the atmospheric oxygen, either directly through burning or through bacterial intermediates, and their carbon-rich fibers are converted back to carbon dioxide and injected once again

*Aerosols are any particles floating in the air, and include dust, smoke, soot, and so on.

into the atmosphere. In 1977 a group of scientists from Rice University and the Universidade de São Paulo suggested that mankind was burning wood and cutting down the forests to such an extent that more carbon dioxide was being released in this manner than was being incorporated by living forests, while another group at the Marine Biological Laboratory in Woods Hole, Massachusetts, claimed that an amount of carbon dioxide equal to the amount contributed by fossil-fuel burning (or even more, up to double that amount) was being thrown into the atmosphere by deforestation.

While those papers were in press, God was having his little joke: the winter in the early months of 1977 was particularly bleak. January of that year was the worst winter month in the United States since records have been kept. There was unusual cold in the Midwest, the Great Lakes region, and the South, including snow in Miami. (Well, that's what the newspaper reports say. I was living in Miami then, and I sure didn't see any snow. Maybe a flake fell before dawn. It was, however, awful cold that winter; my wife even wanted me to turn on the heat in the house, but I couldn't get the pilot light to stay lit.) That same winter saw the warmest-ever January temperatures in Alaska; the springtime plants began to bud there. And there was drought along the West Coast, extending through the Midwest. It was a very weird winter, and we all began to wonder if it had something to do with the carbon dioxide increases. But it didn't. It was due to a phenomenon called Stratospheric Sudden Warming (SSW).

Almost every winter the stratosphere experiences at least one major disturbance that affects the normal circulation patterns, and if this becomes particularly severe the condition develops into an SSW. Very intense high-pressure cells form over the oceans, blocking the normal westerly flow of the jet stream and causing the North Pole to warm, mid-latitudes to cool, and temperatures over the continents to plunge. Until this was explained a year later, the unusual weather had been interpreted as part of the normal pattern, possibly indicating a return to colder temperatures in the future. But now that we understood it, we knew that the cold winter of 1977 should not be fed into the historical scheme to determine future patterns; it wasn't an indication that anything long term was happening, it was just—as Cole Porter wrote—one of those crazy things, one of those bells that now and then rings.

It all went to show how little we know about natural variations in the climate, and thus how little we could trust our models to explain what was happening today, much less to tell us what was going to happen in the future.

We also don't know all that much about the oceans, and the oceans may well turn out to be the key to the whole problem; if not, they are at least a very important component. Basically, the oceans do two things that are relevant: they absorb heat and they absorb carbon dioxide.

In accordance with our practice of simplifying and isolating, we can describe oceanic heat absorption by considering the oceans to be a two-box system. The top box is the surface layer, the water that is in contact with the atmosphere. This layer is well mixed down to a depth of about seventy-five meters, and so as the surface of the water warms up on a hot day the absorbed heat is easily mixed down to that depth. The second box is the deep water, which lies below an effectively impermeable gradient called the *thermocline;* the deep water is thus shielded from atmospheric heat. But in subtropical *gyres,* large water masses in circular motion, which occur in every ocean, the surface waters may mix with the deep waters, and by this mechanism heat can be transported deep into the oceanic depths, where it is stored for—how long? We don't know. The point is that through this mechanism the ocean can act as a gigantic flywheel, absorbing heat from the atmosphere and pumping it down into deep storage. The heat absorbed by the oceans is obviously lost from the atmosphere, which is thus kept cooler than it would otherwise be.

But the oceans cannot absorb heat indefinitely, and the heat they do absorb is not lost from the earth; it is simply stored there, and must someday reemerge. When it does, the atmosphere will be heated more suddenly than we expect; we are sowing the heat, and we may well reap a steam bath.

The oceans are also a sink for carbon dioxide, in two different ways. Physically, carbon dioxide (like all gases) is partially soluble in water, and so some of the atmospheric CO_2 molecules dissolve in the oceanic surface layer. This can be visualized by thinking of the molecules as randomly zooming around the atmosphere, and when their random motion carries them downward against the water, they bury themselves there. But the dissolved CO_2 molecules have the same random motion, and when it carries them up against the water-air interface they can fly out again into the atmosphere. The balance between these two processes determines the carbon dioxide concentrations in both air and water.

And, as was the case with heat, deep motions such as the tropical gyres can carry this carbon dioxide down into the depths and store it there—again, as with the heat, not forever. While it is down there it will increase the dissolving capability of the oceans toward the bottom sediments, which are largely calcium carbonates. This may increase the carbon dioxide concentration of the oceans further, but may also shift equilibrium among different carbonate species in such a way as to increase the capability of the oceans to absorb even more carbon dioxide from the air. Eventually the carbon dioxide content of the oceans may be exceeded, and then we could see a giant burping effect, with an overwhelming amount of carbon dioxide suddenly added to the atmosphere. We have already seen something like this on a smaller scale: Lake Nyos in the African country of Cameroon.

On August 21, 1986, the villagers living around the volcanic lake heard a loud explosion and shortly afterward found it difficult to breathe. The children and old people started to drop first, and then as panic set in the entire village tried to run away. They scrabbled upward along the steep cliffs surrounding the village, but nobody reached the top. Within minutes nearly every one of the sixteen hundred people who had lived in the village was dead.

The culprit was, in a sense, soda water. Our carbonate drinks are made by dissolving carbon dioxide in water, reaching an equilibrium stage at which the dissolved gas spontaneously forms bubbles and degasses when the pressure is reduced, as when we open the bottle or can. All natural waters exposed to the atmosphere have some dissolved carbon dioxide, but not enough to form bubbles—except under a few exceptional circumstances. The Perrier waters are one innocuous example, in which underground charging with CO_2 reaches a level approximating weak soda water. At Lake Nyos a similar thing happened, but a combination of circumstances led to disaster.

The lake surface is warmed by the air, which even in winter keeps the upper waters warmer than the bottom; such an arrangement inhibits vertical mixing. Carbon dioxide is more soluble in cool waters than in warm, and so magmatic gases introduced into the bottom of the lake dissolved readily, building up a CO_2 pressure that was prevented from bubbling out slowly by the pressure of the waters above it. Eventually the pressure built up so high that when the water system was disturbed—which could have been by a small earthquake or volcanic vent, or something as innocuous as a strong wind stroking the surface—the dissolved carbon dioxide suddenly blew out in an immense cloud.

Being heavier than air, the CO_2 didn't dissipate in the atmosphere, but clung to the ground and water surface, pushing the air out of the bowl-like crater in which the village was situated. Carbon dioxide is normally not a deadly gas (we form it in our bodies and breathe it out into people's faces all the time), but in this case it was suddenly present in such a quantity that it took the place of the air, and while it is not poisonous it also isn't capable of doing what oxygen does for us; breathing it in place of oxygen means that the people there suffocated. A brain deprived of oxygen ceases to function within a few minutes, and so the slopes of the volcanic mountain were littered with sixteen hundred bodies within minutes after the lake burped.

This will never happen with the oceans; if their burden of carbon dioxide reaches the burping level, the surrounding conditions are not right for trapping the gas and expelling the oxygen; people living along the ocean shores will not be asphyxiated as the Cameroonians were. But there could be so much CO_2 buried in the oceans that when (and if) they burp, the sudden increase in greenhouse gas might suddenly accelerate the earth's warming to unbearable limits.

The oceans also absorb carbon dioxide biologically, through the creatures that live in the sea, as they build their skeletons and, dying, rain down upon the floor to form sediments, which are buried and eventually shoved deep under the continents. In this manner carbon dioxide is removed from the air, essentially forever as far as we are concerned. (Sedimentary carbonates buried eons ago are now being vented to the surface volcanically, but if the atmosphere increases in CO_2 and the sea creatures absorb this increase, the necessary increase in volcanic carbon dioxide will not occur for millions of years, and this is clearly beyond the time scale we are worrying about. If we haven't learned enough in millions of years to cure our environmental problems, we deserve the problems. Our challenge is to learn enough in the next fifty years to survive.)

Both these effects—absorption and storage of both heat and carbon dioxide from the atmosphere—are qualitatively clear but quantitatively unknown, which is a classy way of saying that we know the effects exist but we aren't able to calculate the extent and importance of either of them. Taken together, they add enormously to the uncertainties in our models.

Lumping all these uncertainties together, then—together with a host of others, of a more arcane nature and limited aspect—what can we say about the predictions of the various models?

At the present time (1989) there are about a half-dozen models regarded as state-of-the-art. All of them predict at least qualitatively the same results on a global basis, but they disagree totally on local geographic effects.

In a sense, of course, this is bad: we all want to know as much as we can about the future. But in a very real sense it is good: whenever the weather changes, it is bad for somebody but good for somebody else. A decade ago we in the United States stopped our attempts to prevent hurricanes, not so much because we were afraid of failing but because we were afraid of succeeding: we realized barely in time that the hurricanes that are a source of worry and damage to us are a necessary source of rain to Mexico, and if we prevent them from forming we would disrupt the agricultural life of that country.

This is true for the greenhouse effect, too. If the world warms considerably, some lands now frozen and barren will become warmer and perhaps wetter; Russia may fare better than the United States. In the world's current state of national selfishness, it might be just as well for us not to be able to predict with any degree of certainty who will benefit from a greenhouse warming, because certainly it will be catastrophic for the world as a whole and just as certainly it is going to take the united efforts of the whole world to avoid this catastrophe.

The models do agree on certain things, although almost never without qualification. Almost everyone accepts that the effective concentration of carbon dioxide (including the effect of other greenhouse gases) will double within the next fifty years, and this will raise global temperatures to some extent. The exact extent is a matter of disagreement, but all agree that it will be in the range 1.5 to 6.5° C.

Unfortunately this is a very wide range, with effects ranging from barely discernible to catastrophic. To put it in perspective, the global temperature has varied by 1° C. fairly routinely, but almost never by more than 2° over the past ten thousand years. A change greater than that would represent the most drastic climatic change that has ever occurred during the entire history of our civilization. The average global temperature at the height of the last ice age, during which so much of the earth's surface was buried beneath glaciers, was just about 5° colder than it is now, so the upper range of the estimated temperature variations puts us right in that ball park. We're not talking cool weather here, we're talking *calamity.*

Well, *possible* calamity. The role of feedback mechanisms, for example, is a source of disagreement and error. Nearly all the current models agree that *if* feedback mechanisms are unimportant, the greenhouse effect won't be so bad: a temperature rise of 1.2° to 1.3° C. is given by all the models, so with reasonable confidence we can say that we have nothing much to worry about if the feedback's don't feed back.

But if they do, all bets are off. The various models account for their effects in different ways, and that is where the majority of the discrepancy—from 1.5° to 6.5° (with a most likely upper value of 4.5°)—arises.

Actually, it's a bit of an overstatement to say that even a 1° rise is unimportant. About five to seven thousand years ago the average global temperature was about 1° C. warmer than it is today, and the effects were certainly noticeable. This is before our history began, so the records are geological and not too detailed, but it is clear that the weather patterns were not what they are now. The tropics had more rainfall than they do today; Africa and India probably had twice as much, and the Sahara was not a desert but a savanna like the treeless grasslands common in Florida. We don't know the full extent of the climate differences, but they were certainly larger than any variations we have seen since we began farming and settling in fixed communities, and so even the minimum temperature rise predicted by the models means adjusting to a world considerably different from the one we have known, and most of the modelers expect that the real effect will be more than the minimum.

Aside from the world simply getting hotter, many complicated differences are predicted. Weather patterns will change, turning fertile areas into desert. Does this imply that deserts will be turned into fertile areas, thus balancing the budget (although at the expense of political stability)? No, it doesn't, because of the rapidity of the expected changes. It is likely that at least some desert areas will suddenly receive a lot of rainfall, but the kicker in that clause is the word *suddenly.* A desert is a fragile thing, particularly when it comes to water. A forest or a grassland has its dirt anchored by the roots of plants; as water falls on it the roots form barriers among which it circulates, slowed down as by a series of roadblocks, allowing it to be absorbed. But in the desert there are no such roots, and most of the suddenly increased rainfall will simply run off, carrying the desert sands with it. Instead of luxuriant life sprouting into being, what we'll see is simply erosion and washoffs.

Sea-level rise is one of the obvious calamities that will accompany a warming world; no model disagrees with this. The only question is how much the waters will rise. If a runaway warming takes place and the West Greenland and Antarctic glaciers melt, the seas will rise hundreds of feet higher than they are today. But even without melting the glaciers, simple thermal expansion of the oceans will raise sea level anywhere from one to five feet within the next fifty years, which makes me very nervous as I type this in Miami—about six inches above sea level. Many of the world's metropolitan centers are built on the edges of the oceans because that was where the world's commercial traffic took place when they were built, and all of these are at risk: New York, Boston, San Francisco, London, Stockholm, Amsterdam. The latter, of course, is *below* sea level, reminding us that the waters can be kept back with dike systems. This can be done, but the cost in money and effort and GNP is a rapidly increasing function of the actual height difference between the cities and the oceans, so once we are convinced that the effect is real and the oceans will rise, we ought to get started as soon as possible to ameliorate the effects.

It's certainly going to be worth our while to do this for the cities listed above, though the cost will be exorbitant. But many of the world's low-lying lands are among the undeveloped nations. The peoples living in the Marshall Islands, many Caribbean and Central American countries, or in Bangladesh are not going to be able to build walls to keep out the waters. We are going to see a refugee problem of global proportions, one that will make the tragic lines of bewildered civilians crossing the war-blasted plains of Europe, or the swarms of Marielitos and Nicaraguans, look like country-day-school picnics.

There are many unexpected consequences to sea-level rise. Not even all developed countries can be maintained with dike systems, regardless of the cost. To take Miami as an example again, there is the problem of the type of land it sits on, which happens to be limestone. This limestone is a porous kind of stone, so that as the water levels rise the increasing pressure will force salt water into the aquifer underlying the land. As we stand behind the dikes with our fingers in the holes, the ocean will seep in underneath us and slowly rise around our ankles, contaminating all the fresh-water supplies as it comes. The entire area will become uninhabitable, and absolutely nothing can be done about it.

Some greenhouse models predict not only long-term effects such as hotter summers, but disturbing short-term effects. Roger Revelle, one of the pioneers in the field, is worried about an increasing incidence and severity of storms. The impact of hurricane-driven water surges all along our coastlines is a terrible threat as the ocean level rises, particularly so since hurricanes gather their energy from the warmth of the oceans, and so will be even stronger in a greenhouse world.

The changed weather patterns will certainly result in a changed river-flow pattern. The current pattern has been essentially unchanged during our history, and our system of nations has grown up around it. If suddenly the agricultural life of a country is wiped out because the river that fed it now flows elsewhere, the political repercussions are inestimable. Both Sudan and Egypt, for example, depend on the Nile. Sudan, lying southward, has been threatening to divert the Nile's waters to benefit its own drought-stricken regions, but Egypt claims a historical right to the natural flow of the river. If the problem should become worse because the Nile waters flow less abundantly, could Sudan allow those waters to flow on to Egypt while its own people die? Could Egypt allow the waters to be diverted?

A rough estimate of the troubles can be gained by considering our own country and some of its water problems. The water-poor western states such as Colorado and Arizona would suffer much the same fate that befell the midwestern dust bowl in the thirties. A 2°

C. warming would result in roughly a 10 percent decrease in rain in the West, which would decrease runoff into rivers by 40 to 75 percent—and today's water requirements would exceed tomorrow's supplies by 20 to 270 percent.

The entire ecosystem, in fact, is at risk. There were some preliminary indications that world food supplies might even increase, since plants grow faster in a higher CO_2 environment, but newer data indicate that though they do grow faster they also grow less nutritious: protein concentrations are reduced. This means that the plants, which in growing faster necessarily need more water (which is likely to be in short supply), cannot feed as many people per pound of plant. In addition, not all plants do as well under changed carbon dioxide conditions; it appears that weeds may fare better than food plants, which means that they would grow faster and be tougher to control, choking off the useful plants. Pests such as locusts are also expected to thrive in the changed climates, tapeworms should do just fine and will probably spread out of the tropics into most "civilized" areas, and—taking one thing with another—the living in that terrible summer will not be easy.

11

A Backward Glance into the Future

We are already irrevocably committed to major global change in the years ahead. . . . The future welfare of human society is to an unknown degree at risk.

—National Research Council, in a white paper to President Bush, 1989

So what then? Are we doomed?

The answers range from yes to no. Dr. Veerauhadran Ramanathan of the University of Chicago believes the effect is unavoidable: "Right now we've committed ourselves to a climatic warming of (at least) 1–3° C, but we haven't (yet) seen the effect." In the summer of 1988 James Hansen at the NASA/Goddard Institute for Space Studies told a Senate committee: "With 99% confidence . . . the greenhouse effect has been detected and is changing our climate now."

And 1988 turned out to be the hottest year on record, as Hansen predicted. But only just barely. The 1988 global temperature was just 0.34° C. higher than the reference period (taken as the average 1950–79 temperature), and 1987 was 0.33° higher. Taking into account the errors of temperature measurements, and the vagaries involved in averaging over the whole globe from a limited number of measurements, 1988 can't be said to be any different from the previous year, contrary to what Hansen and other scientists had predicted. If the greenhouse effect had been responsible for the 1987 heat, then with the continued rise of CO_2 in 1988, that year should have been even hotter.

But it wasn't and many scientists aren't so sure that the present warm temperatures should be blamed on the carbon dioxide. "Climate is a complicated thing," Roger Revelle says, "and the changes seen so far may be due to some other cause we don't yet understand."

Tim Barnett at Scripps and Michael Schlesinger of Oregon State University point out that the variation in warming from place to place does not match the pattern predicted by the models very well, and other groups have shown that the pattern of precipitation does not show the expected changes in equatorial regions.

In May 1989, a gathering of scientists at a Workshop on Greenhouse-Gas-Induced Climatic Change met in Amherst, Massachusetts. *Science* magazine reported that many of James Hansen's colleagues found his views "regrettable," and even "unforgivable. None of the select greenhouse researchers at the meeting could agree with him." They also disagreed with each other about the meaning of many of the research results; about the only thing they did agree on was their press statement: "It is tempting to attribute [the present warming trend] to the increase in greenhouse gases. Because of the natural variation of temperature, however, such an attribution cannot now be made with any degree of confidence."

Yet the Miami *Herald* reported without equivocation on June 25, 1988, that "The greenhouse effect is here." *Time* magazine gave the heat a cover story on October 19, 1987, and a year later the British journal *New Scientist* titled an article "No Escape from the Global Greenhouse." Even the *New York Times* in 1988 stated that "The warming of the earth's climate is no longer in dispute," somewhat begging the question of whether or not that warming is a greenhouse effect or, indeed, necessarily part of a continuing long-range trend.

The media are quick to jump on any sign of potential disaster; after all, that's what sells. Newspapers, television, and magazines heralded the greenhouse effect throughout the hot summer of 1988 as we all sweltered and wondered what was going on. They told us that this was only the beginning, that the hot weather was not "normal" but was the first sign of a new, hotter age.

In August of 1988 my wife and I tried to escape the Miami heat by flying up to the Berkshires, and we found it even hotter there. Everyone was miserable, sweltering, and worried. One afternoon we went to see a play performed at the Edith Wharton house, which is a glorious old, non-airconditioned mansion turned into a museum. The play was well acted, but the heat in the house was unbearable. During intermission we all stood around sweating and mopping our brows, and telling each other Ms. Wharton could never have lived here in such heat, "but of course it wasn't as hot then as now." Later, back in air-conditioned Florida, I found a passage in her autobiography, *A Backward Glance,* where she describes being visited by Henry James in that house during one of the early summers of this century, before anyone had ever heard of the greenhouse effect: "His [Henry James's] bodily surface, already broad, seemed to expand to meet the heat, and his imagination to become a part of his body, so that the one dripped words of distress as the other did moisture. . . . Electric fans, iced drinks and cold baths seemed to give no relief."

Hot weather has always been with us, and not everyone is sure that the recent spate is the first sign of the greenhouse effect: "A few tenths of a percentage point change in the number of droplets per cloud can totally reverse the greenhouse effect," Robert Charlson (University of Washington) has said; and according to Frederick Gadomsky of Pennsylvania State University, "It is much more likely that what we're experiencing right now is part of the climate's natural variability."

"You wait until this winter," concludes S. Fred Singer, chief scientist of the U.S. Department of Transportation. "If we have a real cold winter, you'll see all those guys [new-ice-age prophets] coming back. I'm not joking."

To paraphrase the words Yul Brynner sang in *The King and I,*

> *Sometimes I think the world is going mad,*
> *Sometimes I think things are not so bad . . .*
> *Is—a—puzzlement!*

The problem is that everything is so convoluted and intertwined that it's hard to disentangle the different points of view and the different aspects of the problem that different people are talking about. It will help if we consider separately two distinct sets of data: what we are actually doing to the atmosphere today, and what it is doing to us.

There is no question but that carbon dioxide is a greenhouse gas, and that since the Industrial Revolution we have been pumping increasing amounts of it into the air. The global data were shown in Figure 3; if we extrapolate this trend into the next century, the result is a doubling of CO_2 by 2050, and exponentially increasing concentrations after that. However, this extrapolation is not a prediction; rather, it is a warning, based on a simple mathematical exercise: this is what will happen if we continue to expand our use of fossil fuels as we have done for the past hundred years. (We have simply fit the past data with an equation, and then solved the equation for future years.) Continued increase of carbon dioxide would without doubt mean catastrophe and the end of civilization "as we know it," but this increase is not necessary, and will diminish if we curtail our use of fossil fuels. That's the good news; the bad news is that there are other greenhouse gases in this world, and they are also increasing.

Coal and petroleum products are formed when plants and animals die and are buried by particular geological processes in such a way that they form a pocket of decaying organic material in the earth. Over tens of millions of years these organic compounds are subjected to high pressures and temperatures, and are converted into our "fossil" fuels. Depending on the burial and storage processes, different types of molecular compounds are made from the once-living hydrocarbons, which are comprised mostly of carbon and hydrogen atoms. Carbon atoms by themselves form the solid we call coal, and long chains of carbon atoms festooned with hydrogen atoms form the liquid petroleum products normally called oil and gasoline. When these things are burned the carbon reacts with oxygen from the air to form CO_2.

One other result of the burial process is a gas: methane. We call it "natural gas," although of course any one gas is as natural as any other. Methane is the simplest hydrocarbon molecule, consisting of just one carbon atom and four of hydrogen: CH_4. Compared to the other fossil fuels it has both advantages and disadvantages. Being a gas, it doesn't trap as many impurities as does coal, so it burns a lot cleaner. (As coal forms, deep in the earth, it traps in its structure stuff like sulfur and nitrogen oxides, which upon burning react with water vapor in the atmosphere to form acid rain. It also traps uranium, and so upon burning releases radioactivity to the environment—more, in fact, than nuclear plants of similar power levels do.) Being a gas, methane is more convenient than coal for many purposes; my childhood home was heated by coal, my home today is heated by gas—and I don't have to trudge down to the cellar twice a day to shovel coal as my father did.

Being a gas gives methane some disadvantages, too. We can't use it to run our cars because a gas occupies so much more volume than a liquid that the fuel tanks would have to be as large as locomotives for enough fuel to be carried to get us across town. Further, it is a more effective greenhouse agent than carbon dioxide: it absorbs and traps more infrared radiation (twenty times as much per molecule) and so is a more efficient heat blanket for the earth.

But so what? Buried inside the earth, it isn't a greenhouse problem. And if we pump it out of the earth and then burn it, its carbon is oxidized to CO_2 and this is the form in which it reaches the atmosphere. So it adds to the greenhouse problem by forming carbon dioxide, but in that way it's no worse than the other fossil fuels. In fact it's better, since it releases more energy per carbon molecule than the other fuels do; the more energy we get by burning methane, the less coal and oil we have to burn and so the less of a greenhouse effect we'll have.

The problem is, however, that methane itself is increasing in the atmosphere. This is not owing to our burning of fossil fuels, but to a gallimaufry of other processes. Living in this world is something like watching an Alfred Hitchcock film: just when you think you've got the ending figured out, he—and nature—hit you with another shock. Remember *Psycho*?

In this case the knife-wielding nut turns out to be a bacterium hiding from the air.

First, the data. In 1984, analyses of Greenland and Antarctic glacial ice revealed that methane had been a constant component of the atmosphere throughout prehistoric times, with a concentration of about 0.7 ppm. This condition was maintained until about three to four hundred years ago, when the concentration began to increase, for reasons unknown. In 1987 Donald Blake and Sherry Rowland (the University of California scientist who started the ozone controversy) measured the current methane concentration in the air and found that it amounted to 1.7 ppm. Even worse is the estimate of the current rate of increase, 1 to 2 percent per year, which doesn't sound like too much until you begin to compound it annually and realize it means that while the methane concentration in the atmosphere doubled in the past two hundred years, at 1 percent per year it will double again in just seventy years. Since even at today's concentrations the methane accounts for about one-third of the greenhouse effect, it seems clear that as its concentration rises it may eventually become the number-one villain.

The reason for the increase in global CO_2 is clear; carbon dioxide is the end result of carbon oxidation, the process in fossil fuel burning that provides energy. But the increase in methane was totally unexpected, taking everyone by surprise as the first data came in. We're still not sure exactly where most of it is coming from; about the only thing we are sure of is that it is related to human activities, since historical data show that it has increased proportionately with increases in human population.

But what sort of human activities produce methane? A first guess might be that it's leaking out of the ground in response to our opening up gas and oil wells, but that can't be true since the increase began before we began using gas and oil fuels, and this is confirmed by actual measurements around the drill sites, which put firm upper limits of 30 percent to the amounts that might come from this source. Rather, the most likely sources of the excess methane are the natural producers, stimulated to high levels by human activities.

To understand this, consider *life.* It is a word that cannot be precisely defined, but one thing it is for sure is a system that generates or captures energy and uses it for its own purposes. Plants get their energy directly from the sun, using it by photosynthesis to break down CO_2 molecules absorbed from the air, building their bodies with the carbon and releasing the oxygen back to the atmosphere. Animals do not have this ability; they generate energy by eating plants (and other animals) and oxidizing the carbon in the molecules they eat to CO_2. This is the same process of energy generation that we use when we burn fossil fuels, it just takes place at a slower rate (and therefore lower temperature) in our bodies.

In the early years of the earth, before the plants had become numerous enough to enrich the atmosphere in oxygen, animals could not exist. There were, however, bacteria that were able to get their energy by reducing the carbon they ate to methane in the absence of oxygen; this is not as energy-rich a process as carbon oxidation, but how much energy can a bacterium need?

When the atmosphere became oxygen rich, from the activities of the plants, carbon-oxidizing animals evolved. But some of those early methane-producing bacteria still managed to stick around, wherever they could find an environment poor in oxygen. They need such an environment, since oxygen is poison to them, and they found their environment amid decaying organic molecules in the bottom of swamps and in the digestive tracts of certain animals.

Termites are capable of eating wood, which is indigestible to nearly all living creatures, because in their guts live these bacteria, which chew up the cellulose and change it into more "normal" and digestible carbon compounds for the termites' digestive systems to work on. In doing this, the termites produce, as a waste product, methane, which is then expelled from the termites' guts much as our own guts expel their own waste gases.

Early estimates of methane production by termites were not terribly worrisome. One to ten million tons per year was the range, and though this sounds like a lot, the world is a big place and this would be about one percent of the atmosphere's annual budget. But Pat Zimmerman and a group of co-workers at the National Center for Atmospheric Research established nests of two different termite species in their laboratory and actually measured the methane output. Putting their numbers together with the global termite population—which is not well known, for obvious reasons—they estimated that the termites might be pumping somewhere between 75 and 200 million tons of methane into the air each year.

Part of this greatly increased estimate is due to previous underestimates of termite methane production, and part is due to a suddenly expanding global termite population. The previous production estimates were low because we didn't understand the extent of termite underground nests. In Kenya the NCAR scientists measured methane spewing into the atmosphere as far as fifty yards from any visible termite mound. Investigating more closely, they found a labyrinth of tunnels under the ground, a subway system filled with methane.

Of course, termites have been on this earth longer than we humans, so at first it might seem silly to blame them for the recent increase in methane. But Dr. Zimmerman realized not only that termites produce more methane per termite than had been thought, but that the number of termites world-wide is increasing at a rapid rate. The little pests like dead wood better than live trees, and so the more we kill off the trees the more the termites will flourish. He found that in living tropical forests the termite population is about four thousand per square yard—which is really incredible when you stop to think about it—and that this population nearly doubles when the trees are cut down or burnt off.

Which brings us to the Amazon.

Environmental activists have been protesting the fate of forests all over the world for several decades. Their first argument is theoretical, philosophical: we are merely one of the species of creatures that inhabit this earth, and we have no moral right to usurp all environments for our own purposes. This argument has been opposed by some groups who take literally God's admonishment to Adam, quoted in Genesis: "Be fruitful and multiply, fill the earth and conquer it. Be masters of all living animals on the earth. . . . I give you all the seed-bearing plants that are upon the whole earth, and all the trees."

The environmentalists' second argument avoids this controversy, and is more practical: The world we live in is a complex one, and we don't yet understand how vital each sub-environment is to the quality and indeed the existence of the whole, so we shouldn't go around destroying or even severely changing any of these areas because we might regret it when it's too late.

This latter argument began by focusing on the rain forests' role in breathing oxygen out into the air, and has lately become more directly concerned with their role in absorbing atmospheric carbon dioxide. Now it appears that the most important role the forests play may come when they die and provide an environment for a rapidly expanded termite population and their synergistic digestive bacteria.

The Amazon basin is the world's largest tropical rain forest, and despite the efforts of the activists it—and the world's other rain forests—have been disappearing. Last year the Institute for Space Research at São Paolo reported that thirty thousand square miles had been burned to the ground in 1987 alone, a rate of destruction four times greater than had been thought previously. The burned-out areas were spreading like bubonic plague, leaving in their trail rapidly increasing termite populations to feed on the burnt wood.

The burning of the Amazon is not accidental; rather it is a deliberate policy to clear the land for other purposes, primarily because of a misplaced view of the economics of developing countries. The government of Brazil looked upon deforestation as an inevitable consequence of a growing population in a land of rural poverty. They saw themselves much as the United States saw itself in the last century, as a people with a vast uninhabited territory that could and should be given away to anyone who wanted to settle and exploit it. "Be fruitful and multiply, fill the earth and conquer it."

Sometimes it seems that everything conspires to make this the worst of all possible worlds. Settlers in the Amazon were encouraged to clear the forests by government grants of free land to anyone who wanted to farm or mine it. And to small groups of poor farmers or miners, the easiest way to clear the land is to burn it off. These people do not have the equipment or the manpower to cut down the trees and pull out the stumps; instead they light a match. It's hard to blame them, but the result is that—since fires are easier to start then to stop—vast amounts of land are unavoidably and irresponsibly burnt away. The acreage destroyed by fires and turned over to the termites is many times the land needed or subsequently used, adding greatly to the problem.

Not only that, but the burning itself turns the trees immediately into carbon dioxide and methane. Although the methane produced this way is only about 1 percent of the CO_2 produced, it still amounts to nearly 25 percent of the total from natural inputs and results in the production of forty liters per day for every person in the world.

An ironical element added to the tragedy is that the whole attempt to enrich these jungle regions economically by sacrificing the jungles hasn't worked very well. Export taxes on wood, for example, were rescinded in order to encourage the timber-cutting industry. But the response saw a large number of small operations opening up; these were inexperienced, undermechanized, and terribly inefficient, using up to 50 percent more trees to get a given output of products. And the government did not prosper from the consequent destruction of the forests, because the loss in export taxes was greater than the increase in income from the industry.

(Brazil is not alone in this sort of dumbness. The United States can perhaps be forgiven for its rape of the Western lands in the last century: it was a time, after all, when the world seemed infinite. But today, knowing what we do about the finiteness of the world's natural resources, we continue to auction off timber rights to land deemed unsuitable for logging by the large corporations. The idea is to let smaller companies do their stuff, and to turn a profit to the country.

But the expenses of growing the forests and auctioning off these lands—the costs of surveying and marking, the auctioning itself, and the incredible bureaucratic paperwork involved—is more than the sale of the land brings. If we were simply to let these uneconomic forests stand, let them soak up the carbon dioxide from the air and breathe out more oxygen, let campers and cross-country skiers and hikers and picnickers enjoy them, we would be not only helping the environment but also saving about $100 million a year.)

There are two main causes of the destruction of the Amazon forest. One is an insatiable demand by Japan for wood to be used in construction and paper production. The other is our American appetite for beef, which means a hungry market for the produce of cattle ranches, for which the Brazilian government has been offering tax credits, accelerated depreciation, and other economic hanky-panky in an effort to encourage "useful economic exploitation" of an "underutilized" natural resource. The effort has backfired, costing the government nearly $3 billion instead of turning a profit (although the individuals who own the cattle ranches have become millionaires). And again the United States is not blameless, although this time the blame falls on us, the people, rather than on the bureaucracy of the government, since the biggest single market for all that Brazilian cattle, the driving motivation behind the clearing of the jungle and the establishment of the ranches, is our fast-food industry. All those billions of hamburgers sold by McDonald's and Burger King and Wendy's have to come from somewhere. It seems that no matter how innocent our pleasures, even for one as seemingly innocuous as opening our mouths over a hamburger, we pay in ways unforeseen.

The forests are disappearing not only to make way for McWhoppers, but also to clear the earth for rice fields, and this is a more difficult aspect of the problem. One might think that paddies are the most innocent of human contrivances, but alas it is not so. The methanogenic bacteria evolved on an oxygenless earth, and since then they have sought out niches in the environment where they can get away from our nasty oxygenated air. One of those niches is in the hindgut of termites; another is amid the decaying depths of swamps—and rice paddies. Ralph Cicerone, also of NCAR, led a group making methane measurements in several of the world's paddies a few years ago, and found them to be veritable methane factories. With a rapidly increasing population—indeed, one that seems to be exploding out of all contraceptive control—in many of the Asian countries that depend on rice for most of their nutrition, the acreage devoted to rice cultivation also has been rapidly increasing. The admirable efforts of agricultural scientists to increase the food supply to an endemically undernourished population has succeeded quite well, providing more than one crop a year from many paddies.

All of which also provides more methane to the atmosphere, as much as one hundred million tons per year, a significant portion of the total load and probably as much as leaks out naturally from uncultivated swamps around the world. And pretty nearly as much additional methane is produced by the breaking wind puffed out from the guts of sheep, cattle, goats, camels, horses, and other domesticated animals. Their guts, too, are inhabited by methanogenic bacteria, and their population too is expanding to feed a hungry world. (You and I and the rest of the human race contribute to this problem in a similar way, but by no means as much methane comes out of us when we impolitely vent our gases. The bacteria are present in only about half of us, and the total amount of methane we produce is less than half a million tons per year.)

One further methane worry is more speculative. Under the permafrost of the arctic are molecular compounds called methane hydrates. These are structures in which methane molecules are loosely bound up with water. Should the world get warm enough for the permafrost to melt, these hydrates will be exposed to the atmosphere and will break apart, releasing more methane to the air in another positive-feedback mechanism. We don't know much about this, but the possibility exists of a sudden burp of methane following an uncertain warming, with the result of a sudden and much more serious warming to follow.

Taken together, all the data lead to the inescapable conclusion that the world's greenhouse gases are increasing rapidly, and to the somewhat escapable conclusion that they will not only continue to do so but will increase more and more rapidly as the next few decades roll by and we continue to drive fossil-fuel-guzzling automobiles and munch on our hamburgers (occasionally breaking wind), as we burn off our tropical forests and clear land for rice fields, as we heat our homes in winter and cool them in summer; in short, as we continue to live in the manner to which we have become accustomed.

* * *

If we allow this to happen, if the greenhouse gases do in fact double, the changed atmosphere will in its turn change the entire world we live in. The models have a wide range of errors and disagreements, but most of them agree on the following.

Average global temperatures will increase by about 4° C., with most warming in the northern hemisphere, in the high latitudes, and in the winter. This doesn't seem too bad, since it means warming the places that are uncomfortably cold at the time they are coldest, but it results in changed weather patterns. Most of the models predict about a 10 percent increase in rainfall, generally in the tropics and subtropics, and less rainfall at higher latitudes.

The effect on agriculture would be disastrous; we have all seen how years with warm, dry weather are bad for the crops—and so are years with cool, wet weather. Our agriculture has developed the way it has because that is the best way for our particular weather system, and any deviation means lessened crops. A long-term deviation means we will have to begin using different techniques, and perhaps even go to different crops. The chaos that would result during the transition is easy to visualize for a people that haven't even been able to manage changing to the metric system of measurement, or for whom overproduction of food is sometimes as great a catastrophe as underproduction, resulting in lowered prices that drive farmers to bankruptcy and lead in the longer term to insufficient numbers of farms.

Water supplies will change, deserts will grow, forests will die. Floods and storms will increase in severity and frequency, and the tsetse fly and a host of other tropical pests will "be fruitful and multiply, and inherit the earth."

Living in many now-habitable places will not be pleasant. Several of the models focus on Washington, D.C., as a typical example of our cities that now enjoy reasonable weather most of the time. But a hot day in Washington—a day over 90° F., say—is not at all pleasant. On the average, Washington has about 35 such days a year; after the doubling, it will have 85 days over 90° every year.

The changed weather patterns mean many crops will no longer grow where they now do. They may be able to be grown in different places, but aside from the social problems in moving the farmers, it is not clear that the soils in these new favorable-climate regimes will be suitable for the crops that will now want to grow there.

The ecosystem will be thrown out of whack. Some plants will thrive under an increased carbon dioxide atmosphere and will grow faster and bigger—and may die out because of it, since the animal species that pollinate and distribute them will not have changed their life cycles. If the plants flower at a different rate, the animals will not be synchronized and the entire system will flounder and fail.

The Arctic Ocean, today covered with a two-to-three-meter-thick layer of ice, will become open water. As the warm salt water underneath mixes with the cold melted-ice water, oceanic circulations world-wide will be changed, further affecting the weather in a way that none of the models can yet predict. But in a world balanced as finely as a cyclist on a tightrope, any large change can do nothing but tip it over.

Sea level of course will rise; estimates range from a few feet to hundreds of feet, and even the smaller estimate will be catastrophic for many cities and farmlands worldwide.

In summary, a new world will emerge, one that will put intense pressures on mankind's ability to survive with nature and with himself. If we are unable to cooperate to the extent of preventing the doubling of the greenhouse gases, is it likely that we will be able to cooperate to the extent necessary to ensure the survival of civilization in the changed and disrupted world that will follow?

But of course one must wonder how accurate the models are that predict such terrible effects. Are the results they foresee inevitable? Is there room for error in the assumptions and approximations they make?

Take, for example, the ozone hole.

12

The Hole at the Bottom of the World

The price of liberty is eternal vigilance.
—John Philpot Curran

By 1976 the battle to save the ozone layer seemed to have been won. The scientific establishment had warned the media of the clear and present danger from CFC sprays and refrigerants, the media had warned the public, and the public had cut down their purchases of products containing the CFCs. The billion-dollar industry was crumbling, millions of unnecessary cancers and cataracts had been averted, the good guys had won.

Sure they had.

Early in 1976 Sherry Rowland, together with Mario Molina and a new worker, John Spencer, found some evidence indicating that their earlier estimates of ozone depletion by the CFCs might be 20 or even 30 percent too high. It's always a bit embarrassing to have to announce that your new work makes your earlier work appear wrong, but the Rowland group didn't hesitate. That's show biz, after all. There was an immediate flurry of excitement, and by the next month a group of atmospheric modelers had announced that, following the Rowland group's suggestion, their new calculations showed that the chlorine from CFCs would actually *increase* the total amount of ozone by reacting with nitrogen oxides from natural sources that would otherwise reduce it. There was nothing at all to worry about, they said: spray all you want, the more the better.

In March of 1976 a workshop held in Boulder found that there were errors (or "deviations") of up to 40 percent in the measurement of standard samples done in different laboratories, casting further doubt on the validity of the whole problem. Industry spokesmen stood around on the sidelines and snickered while a lot of academic scientists, embarrassed by what appeared to be an initial overreaction to speculative calculations, began to denigrate the whole affair. (Nothing makes a scientist wince like the word "speculative.")

The London *Observer* reported that "The aerosol scare may be over," and *Science News* headlined, "Rumor and Confusion Follow Ozone-Theory Revision." The New York *Daily News* angrily blasted the scientists for ignoring the uncertainties in their data when they made their initial claims. In May the Council on Atmospheric Sciences—an industry spokesbureau whose purpose was to defend the CFC industry and reassure the public, for which purpose they chose a name that would resemble that of a government organization—announced that the new calculations indicated there was no influence at all by CFCs on the ozone layer; the problem was over.

Some of the atmospheric modelers were saying that 50 or 60 percent of the chlorine released from CFCs was eaten up by the nitrogen oxides; du Pont said it was more like 90 percent. At the September meeting of the American Chemical Society, the two-year anniversary of the meeting at which Rowland had started the whole ozone controversy, Dr. J. P. Jesson of du Pont claimed that if CFC production continued at the present rate for two years, it would result in a depletion of the ozone layer by only 0.05 percent, the effect of which on human health would not be observable. Accordingly, du Pont asked the public and the government not to enact any crippling legislation for two more years, giving the industry that time to complete their studies and to substantiate their conclusions.

The government and the public responded. No legislation was enacted; all pending bills were shelved. And the production and consumption of spray cans began to climb again. In February and March shipments were up by 26 and 41 percent, respectively. Production had dropped 14 percent the previous year, but as 1976 unrolled there was a sense of renewed optimism in the marketplace.

By that summer the pendulum had begun to swing back again. As Rowland and his co-workers investigated further, they came to the conclusion that the new compensating effects weren't as strong as they had first thought; there was a small reduction in the stratospheric chlorine, but not enough to invalidate the original theory or its prediction of a serious depletion of ozone. In September a National Academy of Sciences panel decided that Rowland's group was correct in stating that their first results should not be discarded after all: "fundamentally the theory is right."

Continued growth of CFC production at 10 percent per year, the NAS panel concluded, could not be tolerated: it would result in "major disruption" of the ozone layer within a few decades. Continued release of the spray-can chemicals at the 1973 level would give a 7.5 percent ozone reduction (with limits of error indicating a minimum of 2 and a maximum of 20 percent), an estimate that included the possibility of the Rowland group's new effect. If that effect turned out to be not important after all (and opinions still differed) the ozone depletion would be twice as bad; and a depletion of only a few percent would mean tens of thousands of skin cancers every year.

That wasn't all. Even if the CFCs were to be totally and immediately banned, ozone depletion would inevitably continue for another ten years because of the chemicals already released into the atmosphere which hadn't yet found their way into the stratosphere, and this process would last for another hundred years, and there wasn't a thing we could do about it. The NAS report was a strong condemnation of the CFC industry's stand—but it concluded by suggesting another two years for study before regulation, which was exactly what the industry was asking for.

It's difficult to understand how they could have added that suggestion. It seemed to turn everything else in their report on its head; in fact both the *New York Times* and the New York *Post* wrote stories which gave the impression that the panel had endorsed the industry stand and refuted the Rowland theory. It was obviously a confusing report.

A few days later Russel Peterson, a quondam chemist at du Pont who had quit to become head of New Directions, a private group interested in influencing governments about environmental problems and who was at the time chairman of the President's Council on Environmental Quality, demanded immediate regulation. The National Research Council recommended that CFCs should be banned from aerosol cans by January 1978.

And a new problem was discovered. Not only did the CFCs eat up the ozone, they were dangerous in themselves. Veerauhadran

Ramanathan, who was then at NASA, presented data showing that they absorb infrared radiation at wavelengths of 8 to 13 microns, and therefore must contribute to the greenhouse effect; at the 1973 rate of CFC production, they would give a 0.5° C. temperature increase in fifty years. This was about half the effect being attributed to carbon dioxide; if the CFCs continued to increase in the atmosphere as they had in the past their contribution to the greenhouse effect would equal that of carbon dioxide by the year 2000, and would therefore double the overall effect.

That fall Canada announced it would introduce immediate regulation of the industry. By the spring of 1977 the Food and Drug Administration (FDA), the Environmental Protection Agency (EPA), and the Consumer Product Safety Commission jointly suggested a timetable for phasing out CFC spray cans: manufacturing should be prohibited by October 15, 1978, all companies to stop using existing CFC supplies in aerosols by December 15, and interstate shipments of any product containing CFC propellants to be banned as of April 15, 1979. (There was one little problem: the proposed regulations applied only to "nonessential" uses of aerosols. The problem of defining essentiality was put off till later. Oh, well.)

In December 1977, Sweden became the first European country to ban CFC propellants, effective as of 1979. The ban was opposed by other nations, particularly Finland, which was an exporter of the product. The EEC decided to sit on the fence for a bit, recommending an in-depth review in one year. They couldn't do much else, with the French and British standing steadfast against any immediate action. (This is a problem with countries that value individual liberty rather highly; it often includes the right to do whatever you want even if it hurts other people. You could choke to death on cigarette smoke in an English pub or French bistro, because their smokers have the right to smoke. I guess nobody's perfect.)

In 1979 the National Academy of Sciences doubled its estimate of the expected damage due to ozone depletion by the CFCs, estimating that it would result in a 44 percent increase in the biologically dangerous ultraviolet radiation. To put this into perspective, they noted that there were currently 300,000 to 600,000 cases of skin cancer per year in the United States, resulting in 1,600 deaths: most skin cancers are "good" cancers; that is, curable. But among them there were 13,600 melanomas with 4,000 deaths; melanoma is one of the "bad" cancers. The rate of melanomas was increasing at about 3 percent per year, almost certainly owing to increased sun exposure among the population. A 16 percent reduction in ozone would lead to many thousands of additional cases of skin cancer and a few thousand new melanomas per year.

England's Department of the Environment retorted that the "present understanding is limited and based on model assumptions." The industry agreed, pointing out that despite many years of CFC production and use, no definite depletion of the ozone had ever been actually measured. The response was that confirmation by direct measurements was impossible since the limits of error of the available measuring techniques (taking into account the natural variations within the atmosphere) was 5 to 10 percent, so until that level of depletion was passed no effect could be measured; and the models predicted that level wouldn't be reached until the year 2000—by which time the damage done would be enormous, if preventive steps hadn't already been taken.

The industry three years before had asked for three years to gather more data; now they asked for four years more.

Meanwhile, world production dropped to pre-1973 levels because of the United States' actions, but the rest of world was beginning to take up the slack. Production was expected to grow at 7 percent from 1980 to 2000: the accompanying projected depletion of the ozone was 57 percent! This is so large that the models couldn't even attempt to predict the result: they break down at 30 percent depletion, because such a great depletion changes the input parameters.

Not to worry, the CFC producers reassured us. They now announced that they had come up with a couple of new CFCs that were "safe." Monochlorodifluoromethane ($CHClF_2$) had excellent refrigerant properties, and its production had increased by 25 percent in the last two years. Methyl chloroform (CH_3CCl_3) was proving to be an excellent solvent for the cleaning industry, and its production was doubling every five years.

These two compounds were found to be less inert and so would be partially destroyed in the troposphere, the industry reported. And this was true, but when they were fully tested they turned out to be only *partially* destroyed there; some would get through to the stratosphere. When the data were finally and impartially evaluated it turned out that methyl chloroform at the present levels would contribute nearly half of the total Cl atoms there; it looked as if it might become the biggest CFC problem of them all. But the industry's reassurances, combined with the impression of imprecision in the scientific data, began to convince the public.

By 1983 global production of CFCs again began to increase. In that same year the members of the British Antarctic Survey were beginning to convince themselves that they had found a hole in the ozone.

The British research team had been established in 1957, as part of England's contribution to the world-wide cooperative scientific effort called the International Geophysical Year. Though nobody was particularly interested in ozone back then, the IGY had been established because there was so much about our world that no one knew, and the idea was to measure and try to understand as much as possible. After all, one of the main driving forces of science is the desire to reconcile our ideas with the actual data of the universe, and that can't be done if the data don't exist.

So the British team began to measure the atmospheric ozone concentrations at their base in Halley Bay. At first, during the late 1950s and early 1960s, the data indicated a series of random variations, but by the end of the 1960s there seemed to be a distinctly downward trend to the long-term data. By the time Sherry Rowland began to talk about ozone depletion, in the mid-1970s, the British data looked as if there was indeed a minor depletion, but it couldn't be the predicted Rowland effect because that was thought to be so small it was beyond the limits of the apparatus to measure. Still, no one at the time had a very good idea about the global ozone concentration, and it would be impossible in the future to tell if ozone was decreasing unless we knew what it was in the present. So to establish this baseline kind of data they continued to measure its concentration every year throughout the polar spring and summer. (They couldn't measure it during the polar winter because their apparatus, a spectrophotometer, measured the absorption of sunlight due to ozone in the air; without sunlight the experiment couldn't be done.)

Their data in the late 1970s and early 1980s began to show a distinct, clear lowering of the ozone during the Antarctic spring (around October) every year. But the data were just about at the limits of their precision, so they couldn't be sure. They didn't report their results because they thought they must be wrong, partly because of the imprecision of the data and partly because there wasn't any reason they could think of why the depletion should occur in the spring and then disappear, and mostly because of some impressive high-tech American data that contradicted theirs.

In 1978 the American *Nimbus 7* satellite was rocketed into a circumpolar orbit. Aboard it was an instrument called the TOMS (Total Ozone Mapping Spectrometer) and one called SBUV (Solar Backscatter UltraViolet measurer), which together would measure not only the ozone concentrations but also any increase in ultraviolet light. And for the next several years the *Nimbus* data were regularly published, and showed no such depletion as the Britishers were seeing.

By 1982 the British team was measuring a clear 20 percent depletion of the ozone: it looked as if a "hole" developed in the ozone layer every October. But the published American data did not show any such effect, and it's hard for a team with an old-fashioned instrument of marginal sensitivity to argue against two satellite-flown high-tech instruments like TOMS and SBUV; it's hard for them even to believe their own data, let alone get into a fight over it.

They did manage to convince their administrative purse watchers to okay funds for a more sensitive instrument, and by the Antarctic spring of 1983 they had it delivered and working at Halley Bay. It, too, showed the ozone hole, and it was even bigger that year. In 1984 the hole was up to 30 percent, and another ground station at the Argentine Islands a thousand miles away duplicated the result precisely. With these results they were sure they had something real, and they reported it despite the absence of any similar data from *Nimbus 7.*

Their announcement in 1984 hit the NASA/Goddard Space Flight Center, which operated *Nimbus,* like an earthquake. The Britishers were a solid, conservative group; they wouldn't come out with data like this unless they were sure of it. So why hadn't the *Nimbus* data—oh my God, they realized, remembering suddenly what they had done. They jumped up from their chairs and ran to the computers and pressed a few buttons, and there the data were, right in front of their noses.

When the *Nimbus* was being prepared to be flown with the TOMS and SBUV equipment, there had been a bit of a problem. Not that the equipment wasn't good enough for the job of surveying the global ozone and ultraviolet distributions, but that it was *too* good. It was an automated system, and it would just fly around the world and take the measurements continually, and it would provide too many data for any one man—or group—to analyze.

But of course they had computers to do the analyses. So the data went from the satellite to the computer at Goddard, and the computer analyzed it and presented it to the scientists in organized form. But since there were so many data, they felt (rightly) that they would have their hands full and didn't want to waste time worrying about erroneous measurements. (In any lengthy and complex series of measurements some of them are going to be wrong, because of the many things that can happen to screw up any one particular datum.) So they told the computer in advance to ignore any "anomalous" data.

But how, the computer asked, do I know when any set of data is "anomalous"? When it lies outside normal limits, they said. By the time the satellite was launched, in 1978, there had been many hundreds of ozone measurements made throughout the world, and the limit of variability wasn't very much, something on the order of 10 percent. Anything beyond this range, they instructed the computer, is probably due to a mistake in the measurement and should be considered anomalous; in effect, they told the computer not to bother them with any such data.

It was a silly thing to do, because it meant that if by some unforeseen chance their instruments were to come across something really new and different, the scientists wouldn't be told about it; the computer would reject the data as "anomalous."

And that is exactly what happened. From 1978 till 1984 the *Nimbus 7* actually found the hole in the ozone, but nobody knew about it because the computer didn't tell them. When the British Antarctic ground-based team reported the hole, the Goddard scientists ran back to the computer, which had stored the "anomalous" data, and asked to see it. And there the hole was, clear and evident in the data, just waiting to be looked at.

This is as good a description of the problems with computer models as any. The NASA people didn't look for a springtime hole in the ozone layer because none of the models predicted it. The models didn't predict it because there was an effect up there in the sky that no one had foreseen: icy cloud crystals.

It had been known for some time that chemical reactions can be catalyzed by materials with large surface areas. To understand this concept, visualize a cube of iron one foot on a side. The area of any one surface is one square foot, and with six surfaces the total surface area is six square feet. If the block is not solid iron, but is instead a block of uncompressed sawdust, then each grain of sawdust has its own surface, and the total surface area of the block can be nearly infinitely greater than six square feet. A molecule that is prone to being adsorbed onto a surface would have only six square feet of surface to choose from in the case of the iron block, but it could diffuse into the sawdust block and sit down on the surface of any one of the zillion grains.

In the stratosphere high above the Antarctic, during the long, cold winter, clouds of tiny ice crystals form. Each of these small ice crystals has its own surface area, and on each of these the wandering CFC molecules can be adsorbed. Owing to a phenomenon known as surface catalysis, chemical reactions taking place on the surface of a solid such as the ice crystals are enhanced, speeded up. And so the dissociation of the CFC molecules, resulting in the release of Cl atoms, is accelerated on the surfaces of the ice crystals.

The Cl atoms stick to the surface of the ice grains until the long polar winter is over and the sun returns. When it does, the ice crystals melt and vaporize, releasing the Cl to the air, where they can react with and destroy the ozone. And so every spring the ozone is chewed up, and the "hole" appears. None of the atmospheric models predicted the hole because none of them foresaw the ice-crystal-surface catalysis phenomenon. That is, none of the modelers foresaw it, and so the computers were never told to calculate what would happen if such a thing could occur. Once the hole had been observed, Susan Solomon at NCAR ran back to her computer and told it about the ice-surface effect, and the computer replied with an excellent "prediction" of the springtime hole.

Which tells us two things about the models. In one sense it's very reassuring: the models work when they're given the right inputs. Tell the computers that ice crystals can form in the polar winter and vaporize in the spring, and the computer replies that the probable result is enhanced ozone destruction in the spring. But in another sense it's a bit worrisome: if the computers aren't given the right inputs, they're not going to give us the right outputs. Garbage in, garbage out, as they say. And how many other things are going on in the atmosphere of this complex world that we don't know about and haven't factored into our models?

Unfortunately, it was the second point that people understood best. Or maybe they were just getting tired of hearing about the end of the world, about ozone depletion and skin cancers and cataracts, when they had their own lives to get on with. As warnings proliferated that the Antarctic hole might be a precursor of a global phenomenon that would destroy the ozone layer faster than we had thought, many nations got together at Vienna and signed a Convention for the Protection of the Ozone Layer; but it wasn't enough. CFC production and use world-wide continued to climb.

By 1986 *Newsweek* was reporting that more than 700,000 tons of CFCs were reaching the atmosphere every year. NASA warned that if emissions continued at this rate 5 to 9 percent of the ozone layer would disappear within fifty years. The EPA warned that even a 2.5 percent depletion might mean fifteen thousand additional melanomas per year in the United States alone. The data indicated an even faster depletion: new measurements from *Nimbus 7* showed a 2.5 percent decrease over just the past five years. And in Tasmania, the Australian state closest to the Antarctic ozone hole, the incidence of melanomas has doubled in the past ten years.

But the CFC-related industries were providing 715,000 jobs and producing $28 billion a year, so what are you going to do? The EPA decided not to decide; it would wait until the end of the next year to see if additional regulations were warranted. In Europe the EEC was concerned enough to convene a meeting in Rome to discuss the problem—and several countries sent CFC industry people as their representatives. Nobody was particularly surprised when the meeting couldn't decide to take any immediate action.

In December the United States, Canada, Norway, Sweden, and Finland called for a freeze on all CFCs, and a phaseout to 95 percent below 1986 levels within fifteen years. The EEC and Japan refused to sign the agreement and negotiations broke down, none of the sponsoring countries wanting to handicap their own industrial base unless everyone agreed. The original five countries tried to get everyone to agree at least to ban all "unnecessary" use of CFCs, but the Europeans insisted the stuff made great deodorants, and that was the end of that.

Finally, by 1987 the EEC was forced to recognize the data, and they themselves proposed a freeze followed by a 20 percent reduction within six years, and in April twenty-eight nations met in Geneva and agreed to cut CFC production by 50 percent.

The United States was not one of them. The American-based CFC industry had not wasted its time since the previous December, and now its Alliance for Responsible CFC Policy argued that the most that was necessary was merely a freeze at the 1986 levels. The United Nations Environment Programme countered with a plan involving an immediate freeze, to be followed by a 20 percent cut in 1992 and another 30 percent in 1994—if a majority of the participating nations should agree at that time, based on whatever data might be available then.

Meanwhile, the EPA was predicting literally *millions* of skin cancers and related deaths if nothing was done. The Commerce Department and the Office of U.S. Trade Representative estimated 993,000 American deaths within the next ninety years could be avoided with a 20 percent cutback in CFC production. The cost of those lost lives—in strictly monetary terms, ignoring the human tragedy—would be $1.3 *trillion;* the industrial cost of the phaseout would be less than $4 *billion.* Senator John Chafee (Republican, Rhode Island), head of the Subcommittee on Environmental Pollution, introduced a bill to cut CFC use by 95 percent in six years, because even a 50 percent cut would allow a 5 to 10 percent depletion of ozone by 2050.

But the CFC industry was attacking on all fronts with a new and rather imaginative claim. Ozone in the stratosphere, as we all knew by then, shielded us from the sun's deadly ultraviolet rays; but ozone was also a product of automobile pollution, a major contributor to the deadly smogs of Los Angeles, Denver, New York, and lesser cities. As a pollutant it is a major threat because of its high degree of chemical activity: breathed into the lungs, it eats up the tender tissues there. This pollution ozone, produced at ground level, never makes it into the stratosphere because of its chemical activity; if it doesn't react with our lungs it reacts with almost anything in the atmosphere and is reduced to normal molecular oxygen. But the industry now sought to claim that this pollution ozone, which is increasing every year, would more than make up for any that might be depleted by the CFCs. (And of course they still didn't admit that any ozone at all was destroyed by the CFCs. In March of 1988 du Pont Chairman Richard Heckert told three United States senators: "At the moment, scientific evidence does not point to the need for dramatic CFC emission reductions.")

While the industry used such outlandish claims to fight a delaying action against the banning of CFCs in spray cans (a use that obviously could not be classified as essential and was sure to be abolished sooner or later) they were conducting intensive research aimed at finding other, new uses for their products, uses that would take up the slack when the spray cans went. The Allied-Signal Corporation was mounting a nationwide sales campaign to persuade the electronics industry to use CFCs instead of other solvents. Already a variety of nonpropellant functions had been found and were flourishing: the CFCs were not only integral to air-conditioners and refrigerators, but since the banning of asbestos they had found a home in the building-insulator industry and in the little insulating boxes used for fast-food hamburgers. Despite the aerosol ban, these uses were stimulating CFC production to its highest-ever level world-wide, and the producing nations (the United States, Britain, Finland, Russia, and others) were looking forward to an explosion of orders from the developing world.

Du Pont and the other CFC manufacturers, while busily claiming that there was no "proof" of the harm the CFCs might do, were pushing their chemists to find alternative compounds that would do the same job without endangering the ozone layer. The major contender to emerge was CFC-22, with a hydrogen atom in its molecule that caused it to be not quite so inert as the earlier Freons, and thus to break up in the troposphere and never get into the stratosphere. This compound had actually been developed by du Pont in 1936 but had not made it into the commercial marketplace because it wasn't quite as efficient in many applications.

In the fall of 1988 forty-five nations signed the Montreal Protocol, calling for a 50 percent cut in production by 1999. In an effort to be fair and to put forth a far-reaching agreement, the signees granted permission to the developing nations to increase their use until they caught up to the industrial nations. The overall effect would be a decrease world-wide of 35 percent by 1999. Nearly all the CFC-producing and -using nations signed, with the notable exceptions of Japan and Russia.

So that's pretty good. But is it good enough? Sherry Rowland says the cut in production should have been 95 percent, although our EPA seems happy with the protocol, arguing that without it we would have an additional 131 million skin cancers among people born before 2075. But of course if that is true, how many unnecessary cancers will we have with a 50 percent reduction which could have been avoided with a 95 percent reduction?

In March 1988 du Pont made a sudden turnaround, announcing it would find safe alternatives to their Freon compounds and was planning to phase out their CFCs by the end of the century. The company now urged a complete ban on the CFCs. Their decision was coming none too soon. The latest data showed that global ozone concentrations had dropped several percent over the last decade, which was even faster than the computer models had predicted. Margaret Tolbert, at SRI International in Menlo Park, California, carried out laboratory experiments which showed that droplets of sulfuric acid and water provide a surface for Cl chemistry, much as the icy cloud particles do over the Antarctic; this induces the same rapid ozone loss that resulted in the hole down there. The sulfuric acid in the atmosphere is both a pollution result and a natural component, coming from compounds produced by plants, microbes, fossil fuels, and volcanoes, and so can't be eliminated.

The du Pont turnaround evoked a lot of surprise, but was soon explained. They had found better alternatives than even their CFC-22, and were ready to start producing CFCs -134a, -142b, and -123 immediately. On September 7 they announced plans for a $25 million plant in Texas that would be able to produce one thousand tons each of the new compounds by 1990. In the meantime, the price of CFCs would inevitably rise, providing the company with a $1.8 billion profit, and a headstart on the new alternative propellants.

The change in the industrial heart had nothing to do with a change in scientific data or arguments, but was simply good business. The agreement blowing in the wind at Montreal signaled a change in business conditions, and du Pont decided to jump in and start swimming. "It had nothing to do with whether the environment was being damaged or not," Mustafa Tolba, director-general of UNEP, reported afterward. "It was all who was going to gain an [economic] edge over who; whether du Pont could come out with an advantage over the European companies or not."

But at least the damage has now been contained. On March 1, 1989, the EEC (which produces about 35 percent of the world's CFCs) suddenly decided that the Montreal Protocol wasn't enough, and voted to eliminate all CFCs named in that agreement by the year 2000, and to cut them by 85 percent as soon as possible. The next day President Bush called for America to join the parade. So the war, it seems, has finally been won.*

And what has the delay cost us? The EPA estimates that a 1 percent decrease in the ozone will increase the amount of UV-B reaching the surface of the earth by 2 percent, and this will result in an increase of skin cancer in light-skinned people by 8 percent. Recent data indicates a depletion of 3 percent over the temperate latitudes, causing eighty thousand new cases of skin cancer *per year* in the United States alone, with proportional numbers of cancers all over the world. Other estimates suggest there will probably be a total of a million deaths within the next ninety years in the United States from melanomas if CFC production is finally restricted to 20 percent of the present figures, with only a somewhat smaller number if production actually ceases by the year 2000 (since the CFCs already produced are still in the atmosphere and haven't yet done their dirty work).

It seems like a lot of dead people, simply for the sake of deodorants, hairsprays, and convenient boxes for our takeout hamburgers. But it's nothing compared to the possibilities opened up in the early 1980s by a new computer model of the atmosphere.

"Ladies and gentlemen," came the announcement, "may we present to you the concept of a Nuclear Winter?"

*If we can bring China into line. Early in 1989 China announced that it was planning to raise its CFC production rate to ten times its present level, and by 2000 would be producing as much as the U.S. does today. It's a complex situation. Japan and Italy, who signed the Montreal Protocol, sell China obsolete equipment to make refrigerators, which means the Chinese have to manufacture CFCs to run them. So Japan's and Italy's hands are clean. And so it goes.

13

The Black Cloud

A bunch of the boys were whooping it up in the Malamute saloon. . . .

—Robert Service ("The Shooting of Dan McGrew")

A bunch of the boys were sitting around with their drinks of preference, metaphorically speaking, and complaining about a dust storm on Mars. The problem was that the dust was swirling so thickly they couldn't see the damned planet and their billion-dollar automated space probe was orbiting it helplessly blind.

Carl Sagan was there, together with Richard Turco of R & D Associates, and Brian Toon, Thomas Ackerman, and James Pollack, who all worked at NASA's Ames Research Center in Moffett Field, California. Sagan is basically an astronomer, familiar to everyone with a television set. Turco has a doctorate in electrical engineering and physics, and is an atmospheric scientist interested in what happens when the earth's atmosphere is perturbed by various means. Toon, Ackerman, and Pollack are space scientists who were studying the perturbations induced by small particles in the atmospheres of any planet they could get their figurative hands on.

It really was a bloody shame, they agreed. The government had spent hundreds of millions of dollars and several hundred workers' combined skills for several years to design the vehicle and the experiments that were going to study the Martian surface, and to launch the whole package and guide it across nearly fifty million miles of empty space and slide it delicately into its exact predetermined orbit, from which it could look down for a close-up, detailed picture of the red planet—and when the probe finally got to Mars after nearly a year's traveling through space, they found that Somebody had gone and whipped up a dust storm, which was just whirling and swirling around the whole planet for weeks and weeks, totally obscuring the surface and making their instruments useless, and which never seemed to be going to end. It was enough to drive a strong man to drink.

Instead, it drove them to think of other things. Related things, like dust storms on earth. We don't have such violent, global-sized wind systems here, so we don't have such long-lasting and extended dust storms, but we have had a dust storm or two in our history. There was the volcanic eruption of Tambora in 1815 (chapter 5), and there was Krakatoa in 1883, which provided beautiful sunsets worldwide because of the dispersed and suspended dust. And there was the death of the dinosaurs sixty-five million years ago.

The dinosaurs: now there's an interesting situation, they thought, while their space probe uselessly circled the dusty surface of Mars fifty million miles away. In 1980 Walter and Louis Alvarez seemingly had solved the stubborn puzzle of what caused the dinosaurs' extinction by postulating an asteroidal or cometary impact on earth which raised such a cloud of dust that the sun's light had been shut off, photosynthesis had ground to a halt, the climate had deteriorated, and bang went the dinosaurs and about 75 percent of all species living on earth.

The Alvarez hypothesis had been greeted by a wave of enthusiasm, but also by a reverberating wave of objections. It was currently the subject of one of the most popular scientific arguments, and had transformed the previous debates about dinosaur extinction—and, in fact, about the entire wave of extinctions which had occurred at that time, called the K/T Mass Extinction Event, since it defined the boundary between the Cretaceous and Tertiary geological periods.* Almost everyone had now discarded all the old ideas, such as sudden bursts of cosmic rays from a nearby supernova or a dimming of the sun, and accepted the Alvarezes' idea that an atmospheric disturbance was the cause; but whether that disturbance was a dust cloud generated by the explosion of an impacting asteroid or was perhaps due to a gigantic volcanic eruption was still a matter of intense and occasionally acrimonious debate.

The five frustrated Martian scientists began to wonder if some of the things they had been forced to learn about Martian dust storms

*Yes, I know, it should be C/T, but what are you going to do? The Greek root begins with a *kappa*.

might not apply to terrestrial dust storms. Could they determine if an asteroidal impact or a volcano would be able to generate the dust storm necessary to kill off the dinosaurs?

They set to work to devise a series of equations that would describe what was happening on Mars, and that then could be rearranged and adapted to describe the more massive, more thickly-atmospheric earth. And they failed.

But along the way they began to think of another situation that might throw a lot of dust into the terrestrial atmosphere. Not a volcano, nor a wandering asteroid, but a thick, billowing mushroom cloud.

A solid object thrown into the air will fall to the ground again, and objects of different masses (weights) will fall at the same speed. Galileo demonstrated this by experiment. (Tradition, unsupported by contemporary records, has it that he did this by reaching over the edge of the leaning tower of Pisa and dropping cannonballs of various weights to the ground.) Newton explained Galileo's results with his law of gravity: Each object is *accelerated* toward the earth with a force proportional to its mass, and each object *resists* acceleration with an inertial resistance proportional to its mass, so the masses cancel out of the equations and thus have no effect on the speed of the object's fall.

But of course in the real world all this isn't quite true. Cats, for example, have been known to fall from buildings up to thirty stories high and survive; a block of concrete (or a human being) dropped from that height would shatter (or splatter) on contact. But the falling cat instinctively spreads its paws and tail, and the effect of air resistance slows its fall to a speed low enough to give it a small but finite chance of survival. A human being could do the same, but the cat has a larger ratio of surface area to weight and so the effect of air resistance is greater. Thus the answer to the usually-asked question in Physics 101 ("Which falls faster, a brick or a feather?") is not all that simple: on an airless body like the moon they fall at the same rate, but in the presence of an atmosphere the feather will fall more slowly because of its great surface area.

Objects with a very high ratio of surface area to weight can float in the air for a long time. A leaf, for example, doesn't exactly *fall* to the ground; it *flutters* gently, hanging up on the molecules of air, then slipping to one side and pushing them aside for a short fall, then hanging up again, and so on until finally it hits the ground. Motes of dust are even smaller, lighter, and have larger surface areas; they can float around for days.

The air in our industrial cities is filthy with small dust particles. Take a walk on even a clear day in New York, and when you get back to the hotel and wash your face you find the rinse water in the sink is gray with grime. You have been walking through floating particles of dirt, which stick to you as you make your way through them.

The air after a rainstorm in such cities is clean and bright; everyone has a sense of being cleansed. We say the air has been washed clean, and indeed it has. Rain is an effective scrubbing agent for the atmosphere; the drops of water hit the dust particles and add their own weight to them without increasing the surface area (in fact often smoothing out the irregularities and so reducing the area) and they fall quickly to the ground.

But rain occurs only in the troposphere, and the lower in the troposphere the more frequent the rain. Turco, Toon, Ackerman, Pollack, and Sagan (collectively and acronymously known as TTAPS) realized that the atmospheric lifetime of small particles depends strongly on this rainout effect, and this in turn depends on the height in the atmosphere to which they are thrown. In the first few kilometers of atmosphere the particles will be washed out in a few days; above five kilometers the lifetime increases drastically, to weeks or even months. And if they are lifted all the way into the stratosphere—where rain clouds almost never form—they can float around for a year or more.

And as they thought about this, they saw in their minds the vision of that terrible mushroom cloud rising and billowing over the South Pacific test sites, pushing its dirty, roiling particulates straight up into the stratosphere.

A nuclear explosion is a fearsome thing. Our "normal" strategic nuclear warheads have explosive powers reaching well into the megaton range (having an explosive force equivalent to one million tons of TNT). A one-megaton explosion will dig out a crater with a radius of several hundred yards, many tens of yards deep. Millions of tons of shattered rubble will be thrown into the air, and something like half a million tons of this will be small particles lifted on high.

TTAPS estimated that between ten thousand and thirty thousand tons of tiny dust particles would be injected into the stratosphere *from each individual megaton nuclear explosion.*

How many explosions are there likely to be in a nuclear war? In 1978 the United States Arms Control & Disarmament Agency described our current Strategic Integrated Operating Plan, the overall strategy in the event of war with Russia. Moscow would be hit with 60 nuclear warheads, Leningrad with more than 40, the next 8 largest Russian cities with an average of 13 each. The next 40 would receive an average of 14.4 bombs *per million of population,* and the next *150 cities* would be struck with 25.7 nuclear bombs for each million people living in them. Eighty percent of all cities with a population of more than 25,000 would be struck with at least one nuclear bomb.

And that's just the cities. The Russian military targets, of course, would also be eliminated. There are 1,400 missile silos in Russia, each targeted by at least two nuclear warheads (why take chances?). There are also submarine bases and airfields and storage depots and computer centers and army barracks. . . .

There are a lot of targets out there. And of course *we* have cities and missile silos and airbases, and the Russians have their own thousands of nuclear warheads.

It all adds up to a lot of dust.

And it gets worse.

Let us suggest, as a nonrealistic but illustrative hypothesis, that a cloud of black smoke big enough to circle and enclose the globe could be formed. Obviously, it would block out all sunlight, stop photosynthesis, and end all life on earth. In 1982, at about the time the TTAPS group was working on the scenario they were to call *Nuclear Winter,* an article appeared in *Ambio,* an international journal of environmental sciences published in Sweden, written by Paul Crutzen, the director of the Max Planck Institute for Chemistry in Mainz, West Germany, and by a researcher from the University of Colorado, John W. Birks, who was spending some time in Mainz. They suggested that nuclear explosions not only excavate craters and fling millions of tons of dust into the air, they also burn things. They burn all sorts of things—buildings and trees and people, yes, but also asphalt streets and a host of things not normally thought of as flammable—

simply because of the intense heat generated by the fireball.

Crutzen and Birks suggested that the total amount of smoke generated in a full-scale nuclear war might approximate the black cloud that could choke off all life, and so TTAPS included this possibility in their calculations. They set up a one-dimensional model to test the effects.

A one-dimensional model is not the best sort of model, as we talked about earlier, but the ideas they were discussing were too new and too complex to warrant investigation with a full three-dimensional model. The idea was to take a rough look at the problem and see if there was anything to warrant further investigation.

Actually TTAPS used three separate models. The first was to estimate the amount of smoke and dust generated by a nuclear war, the next was to try to understand what would happen to the smoke and dust thus generated, and the final model was to calculate the effect it would have on the world climate. Each of these steps could be separated from the other, and each involved its own assumptions, approximations, input parameters and calculations.

The results could not be relied on for any distinct predictions of what would happen over New York City or Kansas or Moscow, but rather gave a generalized world-wide picture. The results have been criticized on the basis that we obviously don't have the power to predict tomorrow's weather, so how then can we pretend to predict the weather after such incredible perturbations as nuclear attacks have been added in?

The answer, as TTAPS have written in a letter published in *Scientific American* (January 1985), is that "climate" is not the same as "weather." It is certainly true that we cannot tell with any certainty if it's going to rain in Chicago tomorrow, let alone on any particular day next year, but we *can* tell that next winter in the Midwest is going to be colder than next summer, and that the average temperature in Chicago all next year is going to be lower than that in Miami—and we can predict this not merely from relying on past experience but from mathematical calculations based on physical laws and our understanding of the tilt of the earth's axis and its motion around the sun. In other words, we understand the situation well enough to predict climates, but not well enough to predict the daily variations in climate that we call weather.

Climatic deterioration had been discussed sporadically over the previous decade, as one possible effect of nuclear war; but never had anyone discussed it in any detail, and it had always taken a distant third place in the scheme of things, far behind direct blast effects and radioactive fallout. But now the TTAPS group attempted a serious quantification of the idea, and they found that, along with several side effects of varying degrees of danger, the generation of smoke posed a serious threat to continued human existence.

A one-megaton bomb aimed at a city would be set to explode several thousand feet in the air, so that the blast and fireball effect would not be largely wasted on the dirt beneath the city but would instead be able to spread and cover as large an area as possible. The initial fireball would lash out with a burst of heat that would immediately start a series of isolated fires over an immense circle, some ten miles in radius, with only the most flammable materials being ignited in this stage. Within the next minute the shock wave would spread out over this distance, crumbling and shattering the buildings and, in the process, undoubtedly blowing out many of the initial fires. Secondary fires seeded by the first ones, however, would be spread over the entire area by the accompanying high winds, and over the course of the next few minutes they would merge into several major conflagrations. These, fed by the hurricane-force winds that would rush into the partial vacuum left behind by the passage of the shock wave, would spread and encroach upon each other. The conditions would be worse than those in Hamburg in 1943, the resulting fire storm would suck in ever more oxygen from beyond its boundaries, and all the fires would grow together into one enormous blaze reaching veritably to the heavens as the central updraft would follow the mushroom cloud up to the stratosphere, carrying with it the black smoke and soot that were once the buildings and the people of New York, Boston, San Francisco, or Moscow.

All this is bad enough news for the people who live in American and Russian cities, but the stratospheric-bound smoke is even worse news for everyone on this globe. It is difficult to calculate exactly how bad the news is, since the smoke generated is dependent on so many poorly known factors: on the size of each bomb and the number of bombs, obviously, but also on such parameters as the efficiency with which blast energy is converted into thermal energy and into fires, the type and amount of combustible material in and around each target area, the size distribution and type of smoke and soot particles formed (which are in turn functions of both peak and average temperatures and of material subsumed), and the fraction of that smoke/soot which makes it into the stratosphere.

TTAPS finally estimated that more than 100 million tons of smoke could be generated in a nuclear war involving less than half the total strategic weapons of the two superpowers. Paul Crutzen independently estimated a value of about 300 million tons. This is indicative of the scale of errors involved in the models, and it's important to understand: there are large uncertainties, but that does not mean they are large enough to invalidate the conclusions. TTAPS took the position of using the lowest values for all such estimates, so that their conclusions could be viewed as conservative rather than wide-eyed and irresponsible. As we shall see, their stance didn't save them from that accusation.

One hundred million tons of smoke, distributed as a uniform cloud around the world, would shield us from 95 percent of the sun's light. It would not, of course, be so distributed, at least at first. The smoke would be generated in the Northern Hemisphere, and not homogeneously even there; it would start as thick clouds over the most densely populated regions of America and Russia, but would quickly spread and amalgamate.

If most of the smoke particles are black and sooty, as is expected from the sort of flammable materials found in cities, they will absorb more sunlight than they will scatter away. This means that the cloud will get hot, and in turn will heat the air around it. The heated air will then rise, carrying the cloud with it and extending its atmospheric lifetime. This hot, high-altitude air will form what meteorologists term an *inversion,* a word we are familiar with from stories about the smog situations in cities like Los Angeles and Denver. In the normal situation, with air being heated by radiation from the earth below, warm air is continually being formed at the base of the atmosphere, with colder air above. The warm air rises, the cold air sinks, and we have an efficient vertical circulation system that continually brings in fresh air to the cities below. But when an inversion forms, with cool air below and warm air above, the situation is stable; there is no driving force for circulation, and the city's smog pattern is undisturbed.

In the gigantic nuclear-fueled inversion system we are talking about now, the result would be the same. Vertical circulation would be cut down, the particles would remain suspended high up, and sunlight would be cut off from the depths below. TTAPS estimated that continental temperatures would drop by 30° or 40° C., and that the effect might linger for months or even longer. This gives an average temperature of about −25° C. (or −13° Fahrenheit); and temperature drops of only a few degrees, lasting for only a few days or weeks, raise havoc with most food crops. Daylight would not return for many months—a 95 percent cut in sunlight would give days about as bright as normal moonlit nights—and when the sunshine returned it would be to a dead or dying earth.

The world would be "subjected to prolonged darkness, abnormally low temperatures, violent windstorms, toxic smog and persistent radioactive fallout." Transportation systems would break down, as would power grids, food production, medical care, sewage, and sanitation. Government services would be impossible. All over the world there would be starvation, hypothermia, radiation sickness, disease, and death. "Under some circumstances, a number of biologists and ecologists contend, the extinction of many species of organisms—including the human species—is a real possibility."

There are three additional factors that should be discussed. The first, radioactive fallout, has long been recognized, but the TTAPS study indicates that the long-term exposure of people living beyond the immediate target zones has been underestimated by an order of magnitude.

Radiation damage is measured in units called *rads,* whose definition is too complex for easy discussion, but is a measure of the amount of energy deposited by the radiation per gram of body tissue. A whole-body exposure of 450 rads of gamma rays will kill half the exposed population (considered to be healthy adults; sick people and children die more easily). Doses an order of magnitude less, about 50 rads, will not kill many people directly but will cause cancers and will greatly damage immune systems, causing deaths by secondary effects.

The TTAPS models indicate that millions of square miles near the attack zones will be contaminated with doses of hundreds of rads, and 50-rad doses would cover the entire North Temperate Zone. (Joseph Knox of the Livermore National Laboratory has suggested that if nuclear power plants are specifically targeted, this latter damage zone could have its radiation burden increased to hundreds of rads.)

In addition to the radiation there are the pyrotoxins. It is now commonplace knowledge that most people killed in large-building fires are not burnt to death but are either asphyxiated as the fires devour the oxygen supplies or are poisoned by toxic chemicals released by the burning of building materials. Carbon monoxide (CO) is a major burning product, and is a curious molecule. Its size and structure mimic the oxygen molecule (O_2) so closely that our bodies can't tell the difference; the lungs feed it into the blood supply as if it were oxygen and the blood carries it along to the brain as if it were oxygen, but when it gets there it can't do the job that oxygen does, and the brain dies of oxygen starvation.

There are many other fire-produced poisons (pyrotoxins). The synthetic materials used in large quantities in building materials burn to give hydrogen cyanide and hydrogen chloride. Cyanide compounds were made famous by Agatha Christie and Dorothy Sayers, and most of us remember hydrogen chloride from high school chemistry labs; one whiff of it was enough to burn our nostrils and start us choking. Organic chemicals burn to give a variety of pyrotoxins, similar to the pollutants lofted into the air during traffic jams, which collectively kill tens of thousands of people every year in cities where these conditions are too frequently found.

Finally, we return once again to ozone depletion. In a large-scale nuclear exchange the heat and shock effects of the explosions, combined with the high burning temperatures, will produce large quantities of nitrogen oxides—the NO_x species that were first discussed in connection with ozone depletion by supersonic transports. In our earlier chapters we were worried about depletions by nitrogen oxides and CFCs amounting to a few percent. In a nuclear war enough NO_x could be produced to lower ozone levels by 30 to 50 percent. If anyone was left alive after the slow passage of the nuclear winter, and if these pitiable survivors crawled out of their holes to gaze upward at the emerging sun, their eyes would be burnt out of their heads and their bodies would soon be blackened and burnt to crisps.

14

The Sky Is Falling

It would be folly not to believe that Chicken Little only has to be right once.

—Colonel Thomas H. Magness III, former assistant director for civil works for environmental affairs, U.S. Army Corps of Engineers

And so a bunch of guys sat around calculating *what ifs* and *maybes,* but does their model have any reasonable relation to reality? Do you believe their horror story about the end of the world?

A lot of people don't. In an editorial titled "Reichstag Fire II," the *National Review* compared the nuclear-winter hypothesis to the 1933 blaze set by the Nazis for the purpose of blaming it on communists, thus giving them an excuse to declare a national emergency and secure their hold on the German people. The *Review* suggested that TTAPS had concocted a knowingly false story in order to influence American foreign policy and war-making capabilities. They did more than suggest; they stated that "The idea of nuclear winter was invented in 1982 by anti-nuclear strategists who felt the political need to hype their cause." In a previous article published in that magazine, Jeffrey Hart had "called attention to politically inspired lying on the part of academics and scientists who remained silent—or promulgated data they knew to be false—in the interest of a nuclear freeze (or) stopping SDI . . ."

"Despite the fact that nuclear winter was a fraud from the start, and was widely known to be a fraud in the scientific community," the editorial concluded, "it had a successful run with the general public. The anti-nuclear hucksters largely succeeded in putting one over, few whistles were blown, and the truth of Barnum's maxim was demonstrated once again."

* * *

Barnum is one of those prophets, like Mencken ("No one ever went broke underestimating the intelligence of the American public") who are often honored by both sides of a given controversy. Yes, there is one born every minute; and yes, the lower bound of public intelligence knows no limit; but who's the sucker and who the con man, and who the arbiter of smart and dumb? Who should one believe on a scientific issue, a passel of world-renowned scientists or an obviously political-minded nonscientific journal?

The editors of *National Review* are obviously cognizant of this disturbing suggestion, and have made the necessary effort to garner contrary scientific opinions. They quote an anonymous MIT professor quoting a Soviet scientist who reportedly said to TTAPS, "You guys are fools. You can't use mathematical models like these to model perturbed states of the atmosphere." But aside from the fact that modeling "perturbed states of the atmosphere" is the name of the modeling game, the second-hand anonymous report of a critic does not inspire the greatest degree of confidence in the criticism offered.

So the *Review* does the next best thing: they quote another article that *does* quote eminent authority. In the fall 1986 issue of *The National Interest* Russell Seitz (a visitor at Harvard's Center for International Affairs) "uncovers (the) scientific/academic scandal . . . of nuclear winter." Seitz, who is worried that "Soviet propagandists have seized upon nuclear winter in their efforts to debilitate the political will of the [Western] alliance," provided a direct quote from one of the Western world's greatest and most respected physicists, Freeman Dyson. Dyson's opinion about nuclear winter, according to Seitz, is that "It's an absolutely atrocious piece of science."

Which, coming from Dyson, is damning indeed.

But.

Dyson never said it. In subsequent interviews Dyson has repeatedly denied saying any such thing, denied being interviewed by Seitz, denied knowing anything at all about Seitz or his article. He *has* stated that he is worried about too much attention being focused upon nuclear winter, because if the theory is ever shown to be wrong—and he is not sure if it is right or wrong—the pendulum may swing back the other way and people will begin to think that perhaps nuclear war isn't so bad after all. And that would be tragic.

Seitz is probably not a charlatan, but at the very least there has been a misunderstanding in his article about Dyson's position. The editors of *National Review* are, I think, charlatans in the sense of claiming that the authors of nuclear winter were con men who knowingly perpetrated a fraud, and in the sense of the creationists, who seize upon every doubt in science to throw doubt upon science as a whole and who purposely misdirect their audience's attention to irrelevant facts.

As an illustration of the latter there is an earlier article in the *Review* by Brad Sparks. Titled "The Scandal of Nuclear Winter," it leads off by pointing out that the "Advocates of the 'nuclear winter' theory . . . have neglected to explain something: There was no nuclear winter at Hiroshima or Nagasaki." Which is ridiculous, as the editors easily could have informed themselves by checking with almost any scientist in the world. To imagine that the authors of nuclear winter had never heard of Hiroshima is patent nonsense, and to imagine that—knowing the results of that nuclear explosion—they could formulate a theory that contradicted those results is just as patently nonsensical. Science is, after all, an attempt to explain facts, not to ignore them. The point is that nuclear winter is something that is postulated to occur when *thousands* of *megaton-level* hydrogen-fusion bombs are exploded, while at Hiroshima (and Nagasaki) only one *kiloton*-level uranium-fission bomb was exploded. To compare one with the other is to make the mistake we were all warned about in third grade, not to compare apples with oranges—or in this case, a stack of TNT with a firecracker.

And like the creationists talking about evolution, the *Review* article states that "Nuclear winter isn't science. It is propaganda." Like the Nazis (to whom the editors likened the TTAPS group in their comparison of the Reichstag fire and the winter scenario) they vilify their opponents: "The willingness of prominent men of science to debase themselves and their calling for the cheap thrills of political notoriety is a scandal."

And like both the creationists and the Nazis they simply lie, as when they claim that "Professor Sagan conducted an end run around the ordinary procedures of peer review. In April 1983, he announced his nuclear winter at a closed meeting." Peer review is the standard method scientists use to evaluate new ideas and data. In the ordinary course of things a paper is written and submitted to a scientific journal. The editors of the journal sent it out to three or four other scientists known for their expertise in the field, asking for their opinions. Generally there are at least a few suggestions made, which are either incorporated into the final article or successfully argued against by the authors to the editor, who stands as final judge as to whether or not the article should be printed at all.

This method is an attempt to prevent silly or biased or wrong ideas from publication, and it works pretty well. No reputable scientist would attempt to do "an end run around" it, and the accusation that Sagan did just that is a serious one. It is also demonstrably false. Many, if not most, scientific papers are not only submitted to the peer-review process at journals, but are also presented orally at scientific meetings previous to or simultaneously with journal submission. These oral papers are usually and purposely *not* peer-reviewed, the idea being that ideas which fail the peer-review process are not necessarily bad and should be given some limited exposure to the community. So the charge that a non-peer-reviewed oral paper constitutes an "end run" around normal scientific procedures is nonsense—but it gets even worse.

The "closed meeting" the *Review* alludes to was one at the American Academy of Arts and Sciences, called by Sagan precisely in order that "physical and biological scientists could peer-review the conclusions of the TTAPS authors" previous to publication. The TTAPS group subsequently submitted a short version of their work to *Science* and a more complete version for publication in *Nature,* two of the most severely peer-reviewed journals in the world. (It was published in both places.)

The fact that nuclear winter was in fact extensively peer-reviewed doesn't prove that it's correct, but to claim it's a fraud is the same as the Bible thumpers claiming Darwin was inspired by the Devil. The editors of *National Review* behaved as if they did indeed believe in the Barnum and Mencken dictums. But of course politically motivated criticism does not prove the opposite any more than it proves its own case. After publication, nuclear winter was properly subjected to intense scientific criticism and review (as TTAPS had suggested), and the results were interesting.

The United States Department of Defense was not pleased with the theory, to no one's surprise, for if it were true it meant that our immense stockpile of nuclear weapons was worthless: to use them would be to commit national (and global) suicide. They requested a study by the National Academy of Sciences.

The National Academy immediately conducted a review, and ended by strongly endorsing the TTAPS study. In fact they warned that it didn't go far enough on the dangers of nitrogen-oxide production and subsequent ozone depletion; they told us we should be just as worried about "ultraviolet summer" as "nuclear winter." A conference was called at the National Bureau of Standards laboratory complex in Gaithersburg, Maryland, on Large Scale Fire Phenomenology, which considered both sides of the question. Edward Teller, "Father of the H-Bomb" as well as Father of Star Wars and prime government witness in the Oppenheimer disgrace, suggested that the TTAPS study was an exaggeration of the effects of a nuclear war, citing the large influence of little-known effects. Carl Sagan countered with new data indicating that the group had *underestimated* many such effects. The consensus of the meeting was that the winter scenario was both frightening and possible, and that more research should be done to pin down the uncertainties.

The United Nations Scientific Committee on Problems of the Environment (SCOPE) carried out an exhaustive, two-year research study by several hundred scientists from thirty different countries on both sides of the Iron Curtain. Published at the end of 1985, it concluded that, as *New Scientist* put it, "Nuclear Winter is no fantasy." The Pentagon immediately repudiated the study *on the grounds that it hadn't been peer-reviewed.* Can you believe it? A study by hundreds of scientists from around the world—who could they have sent it to for peer review? Another three or four scientists? In this case the study group *is* the peer group. But who are so blind as those who will not see?

There were many uncertainties in the TTAPS report, as the authors were the first to admit. They had not conceived of it as a final report to the human race on what *would* happen in a nuclear encounter, but as a suggestion of what *might* happen. Before beginning any battle, a competent commander asks his intelligence forces for estimates of all possible levels of potential damage, together with probability levels. Among these will be a most likely result and a worst-case scenario, and both of these are important. If the most likely result is the gaining of ten yards of battleground, with a 50 percent probability, and the worst-case scenario is the loss of the war, with a 20 percent probability, it would be a foolish commander who would sound the charge.

TTAPS, and subsequently the overall opinion of the world-wide scientific community, told us that nuclear winter was not only the worst-case scenario but also the most likely, with a frighteningly high probability if a nuclear conflict between the superpowers was ever begun. They also admitted a large degree of uncertainty, at two major levels.

One level is the matter of input parameters. This includes such factors as the composition of the sooty material lofted into the stratosphere: will the particles be dark and radiation-absorbing or oily and radiation-reflecting? And how high will the smoke rise? What are the physical characteristics of large urban fires? What are the precise mechanisms by which aerosols are lofted into and removed from the highest regions of the atmosphere, and how do these vary with height, winds, or seasons of the year? Are the processes different over land and over oceans?

These seem to be exactly the sort of data amenable to experimental verification but in fact not all of them are, because of the large scale of nuclear warfare. One can study the composition of burning material from forest fires, for example, or the vertical atmospheric distribution of dust resulting from volcanic eruptions, but neither of these scale easily into the quantities, types, or compositions of particles resulting from nuclear explosions—let alone what would come out of not one but thousands of nuclear explosions and the resulting fires.

Much scientific prediction comes from interpolating and extrapolating measured data. Interpolation is relatively easy if you have data that straddle the region of interest, and if the data can be represented by a straight line or at least a simple curve. You can then fit the data points to the curve and read off the values at the interpolated region of interest. If, for example, you have a table giving the average weights of babies from ages zero to five years, and you have a child whose weight was measured from zero to three years and again from four to five years, and the baby's weight lay along the average for those years, you might interpolate the data and with reasonable confidence estimate what his weight was during the unmeasured three-to-four-year period.

Extrapolation is more difficult because it's hard to be sure that the simple nature of the curve continues indefinitely. If the baby gained weight continuously from the age of zero to fourteen, you could draw a smoothly increasing curve through the data. If you then wanted to extrapolate those data to see what the person would weigh at the age of seventy-three, you would come up with an estimate of several thousand pounds, because you would not be foreseeing that the weight gain would stop completely a few years in the future.

Extrapolation from laboratory data to the immense scale of nuclear war is inherently uncertain, and there doesn't seem to be much we can do about that fact because we can't carry out experiments on that scale. We can reduce the uncertainties a good deal by different kinds of measurements, but absolutely precise knowledge is beyond us. Another level of uncertainty, however, and perhaps a greater one, lies in the overly simplified nature of the one-dimensional model TTAPS used.

In 1984 a preliminary three-dimensional model was attempted by Curt Covey of the University of Miami and Starley Thompson and Stephen Schneider of NCAR. Though it was unable to simulate the motion of smoke in response to perturbed atmospheric winds, it was a distinct improvement on the TTAPS model, and it confirmed the basic points of that model: smoke would drastically cool the earth, would change atmospheric circulation patterns, and would spread to other latitudes. One major difference was that the new model did not predict a uniform smoke distribution around the globe, and the resulting patchiness made further predictions of climatic effects more difficult than had at first appeared. "What would actually happen . . . remains unknown." Unknown, but frightening. Both the new and old groups of modelers agreed that further modeling efforts were needed.

In the next few years the NCAR model was improved and two separate three-dimensional models were added to the repertoire, worked out by groups at the Livermore and Los Alamos national laboratories. These models include transport of smoke by winds and removal by rain. Independently formulated, they all agree on the basic effects of a nuclear war: world-wide climatic perturbations, with a long-term drop in continental temperatures that might be close to 10° to 15° C. (19° to 27° Fahrenheit) instead of the 30° originally suggested by TTAPS, with the additional possibility of "quick freezes" in localized but widespread areas.

There are several reasons for the somewhat milder effects of the later studies. For one thing, the oceans have been considered in more detail and found to act as a heat storage unit, modulating the continental cooling. This is probably a real effect. On the other hand, a more thorough washing-out of smoke from the atmosphere may be due to variable input parameters, with the new models injecting smoke at lower altitudes than the TTAPS model. It is not clear which model is more in accord with reality on this point.

The newer results have been referred to by Thompson and Schneider as "nuclear autumn" rather than "winter," and unfortunately this term has been misunderstood by some people. A lot of media coverage emphasized the mistaken notion that the concept of nuclear winter had been "disproved" by the later studies.

Nothing could be further from the truth. As Thompson and Schneider have said, their results indicate that nuclear war would result in "unprecedented and worldwide human misery. . . . [The climatic effects alone] would threaten more people globally than would the direct effects of explosions." Long-term chronic climatic depressions, such as temperature depressions of a few degrees lasting for several years and decreased rain in the subtropics, "which would lead to substantial soil moisture reductions in the summer monsoon belts of the north latitudes" could be "substantially more serious than initially thought," and could initiate a series of reactions that would profoundly disturb and depress levels of civilization worldwide.

A subsequent and more-detailed United Nations report, published in 1988, came to the same conclusions: "The scientific evidence is now conclusive that a major nuclear war would entail the high risk of a global environmental disruption. Residual scientific uncertainties are unlikely to invalidate this conclusion." Later that year the Myrdal Foundation and the Royal Swedish Academy of Sciences invited an international group of scientists to consider the specific issue of the environmental consequences of nuclear war and to evaluate the UN report. Writing in *EOS* (the newspaper of the American Geophysical Union) on January 3, 1989, the conference participants expressed their full support for the findings and conclusions of the UN report. They went further:

"The probability of mass starvation . . . raises . . . the total number of potential casualties from hundreds of millions to *billions.*" Indeed, "The majority of the human population would be at risk." The uncertainties remaining in the calculations, they found, did not make the risk less but more. Combining all these uncertainties, they concluded, *all* nuclear conflict was too dangerous to be acceptable by the human race: "The concept of an environmentally acceptable level of nuclear conflict [is] untenable."

They may be wrong—all the estimates of destruction may be wrong—but if you cling to this hope, you still have to remember Colonel Magness's words: "It would be folly not to realize that Chicken Little only has to be right once."

And the most recent research indicates that not all the early approximations served to make the concept more scary than reality; one at least worked the other way around. The TTAPS group assumed that the soot particles were spherical, even though they knew they wouldn't be, simply because it's easier to calculate the effects that way. But in the summer of 1989 Jenny Nelson, a physicist at the University of Bristol in England, demonstrated that the actual jagged shape of the soot particles means that they won't clump together as neatly as if they were spheres. Thus the total absorptive powers of the soot will be greater than originally calculated: the amount of sunlight absorbed will be greater, the sky will be darker, and nuclear winter will be more deadly than we thought.

And so it goes. To paraphrase a television commercial, it doesn't get any worse than this. The question is, what are we going to do about it?

PART TWO

The Solutions

15

The Legion of the Lost

We're poor little lambs who've lost our way,
Baa! Baa! Baa! . . .
Gentlemen-rankers out on the spree,
Damned from here to Eternity,
God ha' mercy on such as we . . .

—From "Gentlemen-Rankers" by Rudyard Kipling

We are poor little lambs who have lost our way, and if we want to escape being damned from here to eternity we'd better get busy right away finding the correct path. Because right now that light at the end of the tunnel is the fire of hell.

It is a fact that carbon dioxide, methane, and the CFCs are increasing in our atmosphere. It is a fact that they are all greenhouse gases and must necessarily act as does a blanket over a sleeping body: they will trap some of the heat that would otherwise escape. It is a fact that this effect must make the world a warmer place than it would otherwise be. And it is a fact that our civilization is fed by an agricultural system that has become embedded in and accustomed to the climate of the past several centuries, so that a warming of several degrees will bring about a disruption in our food supply that will by comparison make the Irish Potato Famine look like the Seven Fat Years of ancient Egypt. A warming of more than 5° C. will raise sea levels high enough to flood most of our major urban centers. The potential for catastrophe is virtually unlimited. We *must* begin to reduce our production of the greenhouse gases.

In the case of the CFCs, we have already begun and may in fact be at the end of the elimination process, if the accords announced in early 1989 are adhered to, and if China and other developing countries don't upset the rickety applecart. It is notable that our concern about the CFCs arose initially not because of the greenhouse effect but because of the fear of ozone depletion; we didn't realize at the beginning just how effective these gases are as greenhouse agents.

We were lucky. Because it takes time to get rid of something even as marginally useful to civilization as deodorant sprays. The current agreements call for an end to CFC production by the year 2000. If it weren't for the ozone scare we wouldn't be this far along, production would be soaring well into the next century, and the greenhouse situation would be that much closer to ultimate catastrophe.

We can't count on being lucky all the time. And the other greenhouse gases are not as easy to do without.

Carbon dioxide is a necessary result of the oxidation of carbon compounds. There is no way to burn fossil fuels without producing it on a one-for-one basis: each atom of carbon in coal or petroleum produces one molecule of carbon dioxide upon burning. The only alternative is incomplete oxidation to produce carbon monoxide, which is deadly poisonous, or broken-chain hydrocarbons, which are not only poisonous but may also be greenhouse gases. Post-combustion treatment to change these exhaust gases into different, nongaseous chemical compounds is impractical at the present time and for the foreseeable future; it would result in fuel prices too high to be borne and in waste products too massive to be disposed of.

So we must cut down on our use of fossil fuels. That conclusion is absolutely clear and incontrovertible. The questions that remain are: How much? And how?

There are no simple answers to either of those questions. The CO_2 concentration in the atmosphere is currently increasing by several percent every year; if we cut our consumption of fossil fuels in half would we cut that increase in half? To stop the increase entirely would we have to stop using fossil fuels entirely?

No. The atmosphere is only one part of the total global system, and the carbon geochemical cycle continually shuffles carbon between the atmosphere, biosphere, lithosphere (continental crust and underlying mantle) and the oceans. Take the oceans as an example of the complexity of the situation: Carbon dioxide, and all gases, will dissolve to some extent in water. The dissolution percentage will depend on the chemical composition of the ocean (its salinity, concentration of carbonates, concentration of other ionic species), the temperature of the ocean and of the atmosphere, the humidity of the atmosphere, global rainfall patterns, and the concentration of the gas in the atmosphere. Many of these factors, even those well measured in controlled laboratory simulations, are unknown or poorly known in the global environment.

The oceans look like a large pot of salty water, but they are immensely more complicated. They are, for example, excellent buffer solutions, which means that they are resistant to added acids. Acidity is measured by pH, which you probably remember from high school chemistry is the negative logarithm of the hydrogen ion concentration, but if you don't remember what all that means it doesn't matter. The point is that pH values from zero to 7 indicate acid solutions, pH from 7 to 14 indicates basic solutions.

If you add two or three drops of concentrated hydrochloric acid to a liter of water, you change its pH from 7 (originally a neutral solution) to a value of 2.5, which is very strongly acidic. If you take a liter of ocean water (which is slightly basic, with a pH of 8.2) and add the same acid, the pH changes only to a pH of 8. So you couldn't calculate the change in acidity of the oceans simply by figuring how much acid you were adding to how much water, as you can do with river or lake water. This buffering capability is well understood in terms of the concentrations of trace ions in ocean water, which react with and effectively remove the added acid.

In a similar manner, CO_2 added to the oceans can be reacted with and changed to other carbon-containing species, effectively removing the added CO_2 and thereby increasing the ability of the oceans to dissolve even more. But we don't understand this ocean-carbon chemistry as well as we do the ocean's acid-buffering capability. In fact, the entire carbon cycle in the oceans is not well understood. For example, carbon in the form of the several carbonate species ($CO_3{}^{=}$, $HCO_3{}^{-}$, and CO_2) has a measured concentration about three times higher than our best calculations indicate it should be. Why is it so much higher? There are several suggestions floating around, but no answers.

We do understand that the solubility of any gas in any liquid in contact with the atmosphere will increase with the increasing concentration of the gas in the atmosphere. So as the CO_2 concentration in the atmosphere increases, its solubility in the oceans will too. But will the oceans increase their solubility enough to absorb all the added CO_2? No. How much will the solubility increase? How much of the added CO_2 will end up in the oceans and how much in the air? We don't know.

The best estimate of Western scientists is that about 80 percent of added carbon dioxide will stay in the atmosphere, but there is a Russian school of thought that thinks otherwise. V. G. Gorshov claims that the ocean sink for CO_2 is so strong that the ultimate atmospheric increase is limited to 35 to 40 percent, even after we have burnt *all* existing fossil fuels.

K. Ya. Kondratyev, in a recent book introduced and vouched for by Will Kellogg, former associate director of NCAR, estimates on the basis of Russian climate-modeling experiments that the correct value of atmospheric greenhouse warming may be closer to 1.5 to 2° C. than the 4 to 6° most Western scientists are anticipating, largely because of the role of the ocean in the global carbon cycle. He suggests that *at least* 50 percent, and perhaps as much as 70 percent, of the CO_2 released to the atmosphere will end up in the oceans.

As an example of the uncertainties in our knowledge, he points out that when ice melts it changes not only the vertical and horizontal temperature distributions of the ocean it flows into, it also changes its salinity. This can't help but influence—indeed, change drastically—the mixing processes going on in the ocean, and thus its ability to absorb carbon dioxide. These factors have not been thoroughly evaluated; their overall effect is not agreed upon.

And to emphasize our lack of precise knowledge, Kondratyev points out that the CO_2 climate effect predicted by models should have been detectable by surface air temperatures at least in the middle-high northern latitudes by as early as 1980, but it was not. (The mean temperature for 1958–77 was not higher than that for the period 1906–25.) He concludes, in fact, by stating baldly that "There are no grounds for persuasively ascribing the climate warming trend [of the past 100 years] to the impact of an increased CO_2 concentration."

Most Western scientists dispute this view. The surface layer of the oceans is well stirred by winds, and atmospheric CO_2 can easily dissolve and saturate this region, but it is only about seventy-five meters deep and cannot store much carbon dioxide. In order for the oceans to absorb the quantities Gorshov estimates, the CO_2 must penetrate the thermocline, which separates the surface waters from the great oceanic masses below. An experimental approach to this controversy is to trace oceanic water movements through tritium measurements. Tritium is a radioactive isotope of hydrogen, and most of the "natural" tritium in the oceans today was formed in the hydrogen-bomb tests of the 1950s. Its half-life is about twelve years, and so by measuring its concentration in distinct water masses one can measure the "age" of the water mass—or the time since it was last in equilibrium with bomb-tritium in the atmosphere. These measurements, largely carried out at the Tritium Laboratory of the University of Miami, indicate that less than 50 percent of the thermocline region has been penetrated by the surface waters, which contain dissolved atmospheric carbon dioxide. The best estimate coming out of this work is that perhaps 30 to 50 percent of the atmospheric CO_2 might be dissolved in the oceans.

There is the further consideration that the feedback processes are extremely complicated. As the atmosphere gets warmer, the atoms and molecules in it move faster. As they move faster they impact upon the ocean waters with more energy, bury themselves more deeply, and are therefore more efficiently dissolved. But eventually the ocean will also get warmer, and as it gets warmer it expels the CO_2 molecules more efficiently, that is, the solubility decreases; once this process starts the oceans may be a *source* of atmospheric carbon dioxide rather than a sink. The modern oceans contain sixty times the CO_2 of the atmosphere. A 5 percent release of deep-water CO_2 would increase the atmospheric content by 25 percent.

If we are to be honest, we have to consider the oceans a great source of uncertainty rather than a hope of salvation. They are a perfect example of the great scientific uncertainties that confront us. In the face of approaching catastrophe, we are surrounded by these uncertainties; as we approach Armaggedon we find we have lost our way and are stranded in a swamp whose extent we do not know, with patches of quicksand waiting for us in the fog of our ignorance.

What do we have to do? Before we can do anything else we must begin to dispel this fog: we have to learn all the things we don't know. We have to understand the basic chemistry and physics of the oceans and the atmosphere before we can attempt to change them to our benefit; we must learn atmospheric evolution to see how the air got to be the way it is, in order to understand better how our activities may change the way it will become; we have to know what is going on—and we don't, not yet, not by a long shot. We need to spend more money and direct more attention to basic research efforts, and we're not doing that. When President Reagan took over, for example, the military research and development budget jumped from $13.9 billion in fiscal year 1981 to $21.5 billion in 1982; in the next two years it grew to more than half the entire federal research funding. Increases in other areas dropped to below inflation rises, with energy programs suffering the deepest cuts.

It hasn't been getting any better. In fiscal year 1989 the National Science Foundation budget for atmospheric chemistry research was not increased one penny over the previous year, despite the inroads of inflation. In effect, the research on the basic chemistry of our atmosphere was cut back at the very time when we stand on the threshold of disturbing it to the point of our own extinction precisely because we don't understand it well enough!

So the first thing we have to do is get our priorities straight. What is the greatest threat our country faces? An attack by the Russians? Or the destruction of the environment at our own hands? As Eisenhower said, "The problem with defense spending in a democracy is trying to figure out when you are destroying from within what you are trying to protect from without."

It's important to understand that the necessary research is in large part basic rather than applied. There is a tendency to treat science as a business, to say that we're willing to spend money but we want value received for monies expended: a decent bang for the buck.

It's a good concept; it seems to make sense. But it doesn't work.

It doesn't work because, unlike a normal business run on sound commercial lines, science is messy and disorganized, a jumble of the unknown being disentangled by the immature. Sorry, but that's the way it works.

Let me give you an example. The standard case story of the success of science in coming to our aid is the Manhattan Project. We were engaged in a terrible war, and a few scientists told the government that they might be able to work out a method of generating energy in a quantity so enormous that one bomb would end the war. The government said to them, "Here's all the money you want. Go do it." And they took the money, and they did it.

But it's not that simple. The success of the project depended on many bits of basic ideas that had been developed previously, with no such end as a bomb in mind. Take two particular points. First, the basic idea behind all nuclear energy is Einstein's equation $E = mc^2$, which told us that nuclear mass could be rearranged to yield an immense amount of energy. That equation was developed in 1905, as a result of Einstein's attempt to understand a totally unrelated aspect of the universe: the nature of light. He failed in his attempt, but along the way he developed relativity, which led in turn to our understanding the radioactive energy discovered by Marie Curie a few years before.

Had Marie Curie not done her experiments—which were undertaken only to understand a rather weird observation of Henri Becquerel's, and which seemed of little importance to anyone at the time—and had Einstein not embarked on his studies that led to relativity, what would the scientists have answered in the early 1940s when they asked themselves if they could find and exploit a new source of energy? They would have had to answer, "There is no such energy source. It doesn't matter how much money the Government throws at us, or how urgent the needs of society, we don't know where to start."

When Michael Faraday in 1821 discovered how to generate minute quantities of electricity, just barely enough to make the needle on his measuring instrument quiver, and demonstrated his new apparatus to the Prime Minister of England, he was greeted with the observation "Very interesting, of course. But of what use is it?"

To which Faraday replied honestly and presciently, "I don't know. But I'll wager some day you'll be taxing it."

To solve a specific problem, you need to direct resources at it, you need applied research. But if the body of basic research hasn't been done already, to give you at least a general understanding of that aspect of the universe you want to influence, you haven't a chance, you're a dead duck. And currently our basic understanding of this particular aspect of the universe is sadly lacking. We don't understand the oceans: we don't know how the carbon cycle in the oceans will react with that in the atmosphere, nor how heat will be transported and stored in or released from the oceanic environment. We don't understand clouds: we don't know how they will react to an increased temperature and whether they will trap heat from below or reflect away heat from above. We simply don't understand the earth well enough to begin fiddling with it; we're like a child who has been given a screwdriver and proceeds to begin unscrewing the electric light switch to see what makes it work, and whose next step will surely be to stick the screwdriver in among the exposed wires.

And in the meantime we spend billions of dollars to design nuclear-pumped x-ray lasers to destroy incoming missiles, and detection systems for enemy submarines, and antisatellite rockets. . . .

I think we're strangling from within what we're trying to protect from without.

It is not easy to accept the idea that when faced with a staggering problem of immediate importance the first thing we must do is spend money on basic science, which may or may not impact upon the problem at hand. We are justifiably proud of "American know-how" and want to apply it in a businesslike way to solve our problems; but "know-how" comes *after* "know-about," and we don't yet know enough about this world we live in.

I'd like to use an analogy suggested by Richard Feynman, and take it a few steps further. Feynman liked to compare scientists doing basic research to spectators with obstructed-view seats at a celestial chess game. The spectator scientists, who know nothing about the game and who can see only part of the board, are trying to figure out the rules; this would be tantamount to "understanding the laws of the universe," for in this analogy the chess game is the universe and the rules are the laws of nature. (Eventually the spectators might even hope to be able to deduce the personalities of the gods who are playing: this would be the final unification of religion with science.)

The obvious way to begin the task is by observing as much as possible of the game; and so in science the first thing we do is make observations of the workings of nature. Since we can see only part of the board, it's obviously very confusing. We see pieces coming and going, and at first there doesn't seem to be any pattern. But then we might notice, for example, that in some games a black bishop is taken in our part of the board; it is removed. In some of these games another black bishop will later come into our view, so one of the rules must be that there are at least two bishops per side. If this bishop is also taken, we don't see another one appear, and we might conclude from this that there are only two bishops. But since we see only part of the board, how can we be sure that there isn't another bishop somewhere else? We can't, but as we watch more and more games and see over and over again that if one bishop is taken another may appear but if two are taken no more appear, we finally agree at least provisionally on another law of nature: only two bishops to a side.

Next we notice that when a bishop is taken from a red square, the other bishop, when it appears, will always be on a black square. And for the remainder of the game that bishop will remain on the black squares. No other player in the game remains color bound like this, so the observation is striking. We conclude that each bishop in the game is set on a different-color square and remains with that color forever. This is clearly a fundamental law of nature, but one that is not understandable.

Until we make the further observation—which is so simple that we should have seen it earlier—that the bishop in our small section of the board always moves in a diagonal, and now we understand why it is constrained to remain on one color only. With this discovery we have replaced one fundamental law—the one-color law for bishops—with a more fundamental law that explains it, the law of diagonal movement. We now understand one of the rules of the game.

Or we think we do. Because some scientist, always testing our supposed understanding, now blows us away with the discovery that sometimes—not very often—hardly ever, in fact, but *sometimes* when there is only one bishop in the game it will change colors! How is this possible? No one understands, and all our understanding of science is thrown open to question. (What has happened is that in another section of the board, hidden from us, a pawn has advanced all the way and has been changed into a bishop, with a fifty-fifty chance of being either red or black, independent of the color of the bishop we have observed.) When the original bishop leaves our field of view and the new one occurs on a different-colored square, we naturally think it's the original one that has changed colors. Eventually, if we try hard enough and observe long enough, we will understand what has happened, and then we will have a new understanding of the game.

And so, little by little, we gain some knowledge of the universe in which we live. At the present stage of our civilization we are just about as described above: we have some knowledge of the workings of the bishops, but we can't claim to understand the game. We certainly don't understand knights, for example, nor do we even understand the object of the game, since in the games we have so far observed the taking of the king has always occurred elsewhere on the board, beyond our observations.

What is frightening is that somehow we have learned to manipulate some of the pieces. We are changing the game, a game that we understand very incompletely. We don't know what we're doing, but we're doing it. And the gods are likely to get very annoyed with our kibitzing and nudging of the pieces, and may say, "To hell with it," and upset the board and walk away, leaving us to wonder what happened.

In the early 1950s President Eisenhower met with his cabinet and asked them to give serious consideration to whether or not the United States should launch an unprovoked nuclear strike against the Soviet Union. He wondered "if our duty to future generations did not require us to initiate war at the most propitious moment we could designate." He never allowed that question—or any hint that he was considering it—to reach the American public. "We don't want to scare the country to death," he said. But he was considering the question.

If the answer had been in the affirmative, we would have hit Russia with fifteen hundred nuclear bombs. Moscow would have been struck with twenty twenty-megaton bombs and Leningrad with twelve. "All of Russia would be nothing but a smoking, radiating ruin at the end of two hours," promised General Curtis LeMay, our SAC Chief.

And what would the United States have looked like at the end of a year? According to what we now know—but hadn't the faintest notion of then—with such an attack we could have brought a nuclear winter down on our heads. No one in Eisenhower's cabinet thought to ask about the possibility, *because it hadn't occurred to anyone.* Nuclear winter was conceived thirty years later by a bunch of scientists who had been thinking about dust storms on Mars; if they hadn't been trying to understand the solar system they would never have come up with the idea. And that's the point about basic research: without it, you don't even know what questions to ask.

So. Priority Number One: We must accelerate our efforts to understand the basic chemistry and physics of the oceans and the atmosphere. Unfortunately, this costs money; and unfortunately it isn't enough to solve the problem, it's only a beginning.

But fortunately, if we tackle Priority Number Two at the same time we can actually save some money.

16

The New Conservatism

Without improved efficiency, it is only a question of which will collapse first: the global economy or its ecological support system.

—Christopher Flavin and Alan Durning, Worldwatch Institute

We should probably renumber this Priority 1A instead of Priority 2, because we must begin it right away rather than waiting until we understand more about the atmosphere. We must—and can—cut down significantly on our use of fossil fuels simply by conservatism.

Back in the 1950s the political conservatives drove around in Buicks and Cadillacs, while the young wild-eyed liberals made a lot of ugly noise with their tinny little Volkswagens; but today the true conservatives are those driving small, fuel-efficient cars. Half the oil we burn in this country is burnt in cars, and the efficiency ranges from about twelve miles per gallon for Jaguars and Rolls-Royces to upward of fifty miles per gallon for Hondas and Escorts. Clearly we could easily cut the fuel we burn on the roads by half, cutting our total oil budget by a quarter, if we insisted on driving the right kind of cars.

And yet by and large we don't. When the gas crunch hit in 1973–74 and again in 1979–80, we panicked at our obvious dependence on foreign oil and, among other reactions, enacted legislation insisting that American gas guzzlers be made more fuel efficient. But when the crunch eased and the long lines at the gas stations disappeared, so did our anxiety. Despite the fact that the danger we are now facing is much more fearsome than the inconvenience of gas-station lines, in 1986 we permitted the government to roll back the fuel-efficiency requirements for new cars from 27.5 to 26 miles per gallon. This small cutback results in a waste of fuel every year greater than that which could be provided by both the offshore California oil fields and the off-limits fields in Alaska *combined.* This waste of fuel is not only a waste of money, but contributes to the wasting of our climate: every drop burnt in our cars provides more carbon dioxide to our air. Amory B. and L. Hunter Lovins in the December 1987 issue of the *Atlantic* make this conservation argument strongly, and although their concern is avoiding another oil crisis rather than preserving our climate, their solution is a necessary part of our solution. As they point out, the actual efficiency increase in cars used in America between 1973 and 1986 was from thirteen to eighteen miles per gallon, and this saved *twice* as much oil as we imported from the Persian Gulf during that same period.

We obviously have a case where American national security can coexist comfortably with the preservation of the global environment. And there's a lot more we can do. In 1989 the city of Los Angeles installed a new system for synchronizing stoplights, based on the National Institute of Standards and Technology's atomic clock in Boulder, Colorado. The usual system often breaks down; in L.A. more than a dozen stoplights a day used to go out of synch. The new system, which uses radio transmissions rather than error-prone telephone lines, will enable motorists to cruise through town with fewer stops, which means fewer cars sitting idle with their engines turning over and burning gasoline to no purpose. Aside from the improvement in traffic flow and motorists' nerves and pollution-caused diseases, this will save twenty-two million gallons of gasoline every year in just that one city.

Twenty-two million gallons of carbon that is not oxidized to carbon dioxide will not solve the greenhouse problem. It's a drop in the bucket. But with enough drops, the bucket will fill. There are a lot of cities out there with long lines of automobiles waiting at stoplights, pumping out the residues of wasted fuel into our air.

And there's more that can be done, much more. Building insulation is one particular field that beggars the imagination. A test home built in New York with improved insulating efficiency, at an increased cost of only 5 percent, is being heated by an annual energy expenditure equal to that of most American refrigerators! Part of that is due to the high inefficiency of our refrigerators, which currently consume about 7 percent of our total electricity; but that story is beginning to change. Models on sale now use nearly half as much energy as those sold fifteen years ago.

Other aspects of energy conservation are not nearly so encouraging. Federal funding for energy-conservation projects fell from $344 million in 1980 to less than half that value in 1987. It seems that conservatives don't like to conserve, even though the indications are that simply applying those measures we already know about could save our economy about $100 billion a year.

Some of these are relatively simple and certainly will be coming along soon. We now spend nearly 25 percent of our electricity on lighting. New devices that will soon be available on fluorescent lights will improve their efficiency 30 percent, which can save more than five hundred million barrels of oil in the next twenty years. A new discovery, not yet ready for commercial exploitation, holds the promise of improving the efficiency a further 100 percent by providing two photons of light instead of one for each unit of radiation generated.

More than 50 percent of our electricity is spent to drive motors, and almost all of these run at fixed speeds. Almost half that energy could be saved by simply running them through power-electronics systems that would enable the motors to run slower when their load is lighter.

But it is in the home that most savings are possible, largely because we have been so inefficient in the past. A third of the heat lost from American homes escapes through closed windows, but we now have the technology to cut down such loss by 90 percent. We have high-efficiency water heaters available now, but many people won't buy them because they cost slightly more and they don't realize how quickly they'll earn back that money on energy savings. And every penny saved by energy conservation is equivalent to less carbon dioxide added to our atmosphere.

It's also equivalent to money in the bank. Professor John Blackburn of Duke University estimates that we waste fully half the energy we generate in America. The data support him: we spend 11 percent of our gross national product on energy, while Japan spends only 5 percent. This translates to about a 6 percent economic edge on everything they sell. The wastage is high enough to be compared to the energy production of major oil fields. Lovins and Lovins point out that if we spent as much on insulation for our buildings as we now spend on military forces to guard the Middle Eastern oil fields, we wouldn't *need* those oil fields; we could become energy self-sufficient. And aside from our national security interests that would be served by such independence, isn't every drop of fuel *not* used equivalent to that much of a greenhouse gas not added to our atmosphere?

Well, to be absolutely and painfully honest, no. Sorry about that; it would be so much simpler if it were. But as things stand today, the insulation used in the housing industry will probably use CFCs as foaming agents, and when the insulation eventually decays it will release the CFCs to the atmosphere. So there is a down side to all this, but when one crunches the numbers they still come out overwhelmingly on the side of conservation. So even if things aren't as simple as they might be, the direction all the signs point to is clear: conservation of our fuel supplies can lead to real, distinct, and measurable improvements in the greenhouse situation. *And* at a real savings in money and an improvement in our national security.

We're already doing this. From 1977 to 1985 the United States gross national product grew 21 percent and the number of trucks and automobiles grew 20 percent, but our oil consumption went down by 15 percent. "The oil savings in 1985 alone equaled three times our 1986 imports from the Persian Gulf," according to Lovins and Lovins.

On the other hand, it's not always easy to know what to do. In March of 1977 the *Science* printing plant ran out of gas to dry the coating of the slick paper the journal customarily uses, so they printed that issue on nonglossy paper. When they apologized in print, a score of scientist readers wrote in to say they should always print on that paper just to save gas and conserve energy. But because without the coating the pages tear easily, they had to use thicker paper and so consumed an extra sixteen tons of wood pulp, requiring that more trees be cut down. And global deforestation is another part of the problem.

We've already seen how, when trees are burnt, the carbon in them is oxidized to CO_2 and released directly into the atmosphere. Deforestation is currently releasing one to three billion tons into the air every year. This is a significant amount of the total atmospheric budget, so if we could put an immediate "cease-fire" to deforestation, the greenhouse problem would vanish overnight, at least temporarily.

In addition to the direct combustion of organic carbon to CO_2, deforestation destroys a major atmospheric carbon dioxide sink, for trees—and all living plants—breathe in carbon dioxide and breathe out oxygen. The carbon dioxide they take in is incorporated into their bodies and stored there, and is thus removed from the atmosphere. Two million square kilometers of forest can store one billion tons of carbon per year for about forty or fifty years. When the plants die they normally decay and some of the carbon is returned to the atmosphere, but a large portion is buried in the soil: the total amount of carbon in a forest exceeds substantially the amount in trees at any given time. Eventually this carbon will be returned to the atmosphere, but "eventually" is a long time; and time is part of the problem.

The impending greenhouse warming has the potential to become a catastrophe not only because the earth will become warmer than it has ever been before, but because the change will take place more than ten times faster than it ever has before. (At the end of the last ice age, ten to fourteen thousand years ago, the earth warmed 3° to 5° C. over a time period of a few thousand years; the current greenhouse effect may result in a similar warming in less than fifty years.) The ecosystem of the world is a responsive system, changing as the environment changes, but within tight limits. We—the entire biosphere—can learn to live in different conditions than those we are used to, but only if we are given time to make the change; the anthropologically driven greenhouse changes are expected to come too quickly for many of our vital systems to be able to accept. The climate of Maine could become like Georgia's is today, within the lifetime of a single tree. The vegetation growing in Maine might be able to migrate northward, and the southern vegetation might expand into Maine, but not within one generation. So if we can slow the temperature increase by keeping some of the carbon tied up in forests for tens or even hundreds of years, that has to help.

But of course there would be a large social disorganization if we were to insist that trees must replace crops, which brings us to the third component of the impact of deforestation on the environment. This is related to the *manner* in which we are losing the trees; many of them have been burnt away to clear land for human activities, and as we have seen this provides lovely feeding grounds for an expanding termite population, which in turn provides a breeding environment for intestinal bacteria, which generate methane and expel it to the atmosphere.

This loss of forest lands is serious throughout the developing tropical and subtropical world. Most of the rice fields around the globe were dry tropical forests before the advent of our own expanding population; instead of being a sink for carbon dioxide they have become a source of methane. The tropical forest in Costa Rica has shrunk from 20 percent to 2 percent of the area in the past twenty years, as the people have burnt it off to clear it for crop and pasturelands.

We can hardly expect already-undernourished people to turn their paddies and farms back into forests so the summer weather in Washington doesn't get too muggy; the "immediate cease-fire" referred to above is not going to happen. But other things are happening, and each of them has an impact.

Applied Energy Services is an American company with three small coal power plants in the eastern United States; it has two more under construction and is planning three more. It has started a cooperative project with CARE: over the next ten years CARE will help forty thousand farmers to plant fifty-two million trees in Guatemala, to reverse some of the impact of the lost forests there. The project will remove fifteen million tons of carbon from the atmosphere, which not coincidentally is the amount AES will emit in its new 180-megawatt plant in Uncasville, Connecticut, in its forty-year operating life. AES is giving $2 million to the project (about 1 percent of the total cost of the power plant), with CARE and the United States Agency for International Development each giving the same, and the Guatemala government is providing $1.2 million. The project will also use the services and training of Peace Corps volunteers worth an additional $7.5 million. So it's not cheap, and it won't solve the world's problems: it would take ten million acres of trees (an area twice the size of New Jersey) to offset ten years' worth of emissions from new American plants alone, but it's a start.

To absorb five billion tons of carbon (the amount entering the atmosphere every year) would take seven million square kilometers of newly planted forest. That is the size of Australia. And we would have to plant this amount of new forest every year to solve the problem created by all CO_2 sources. Obviously we're not going to do this. But this is also just about how much tropical forestland has been cleared for agriculture, which gives you a pretty good idea of the impact of deforestation.

So while recreating the forests is not a practical answer to the greenhouse effect, it can contribute to the solution, and a good number of small solutions may turn out to be just what we need.

In 1988 the American Forestry Association planted one willow oak tree near the White House. That's not going to help much, but it was only the beginning. When completed, the Global Re-Leaf Project will have planted 100 million trees throughout the United States. These trees will remove 18 million tons of carbon from the air. That amount is less than 1 percent of the problem, but if most of these trees are planted in urban areas, the shade they provide could reduce air-conditioning requirements by $4 billion a year, which is 6 percent of the entire U.S. residential electricity consumption. This would save the combustion of 16 billion tons of coal, preventing 60 million tons of CO_2 from entering the atmosphere, which is about 2 percent of the annual accumulation. These small efforts do begin to add up. (We could do even more. At the present time the United States Department of Agriculture is setting aside less than 25 percent of highly erodible cropland under their Conservation Reserve Program. To use this land for reforestation would cost seventy cents for each ton of carbon removed from the atmosphere. Planting urban shade trees would cost less than ten cents for each ton.)

There are even smaller efforts, which shouldn't be denigrated, because as the man says, "If you're not part of the solution, you're part of the problem." Students for the Children's Rainforest, an organization started by Sharon Kinsman of Bates College in Maine and supported by local organizations such as the Florida Rainforest Alliance, is an uprising of elementary through high school students and teachers. They collect and recycle paper, preventing the destruction of more trees, and use the proceeds to buy and preserve forest land in Costa Rica. And they talk to other kids and their parents and local civic organizations to tell them about the problem and enlist their help.

More formally, we have several legislative bills pending in Congress. Senator Timothy E. Wirth (Democrat of Colorado) has introduced the National Energy Policy Bill, which would provide funds to monitor tropical forests and develop reforestation plans worldwide, and Representative Claudine Schneider's (Republican of Rhode Island) Global Warming Prevention Bill calls for us to lead the way in obtaining an international agreement to cut atmospheric carbon dioxide concentrations by the end of the century. And it's more than hot words: domestically, it would provide hundreds of millions of dollars for research into alternative energy sources; internationally, it will offer financial aid to developing countries for family planning services, to cut the population growth that is forcing the need for more and more energy.

We are already seeing some response to these initiatives. In Brazil, the government not only decreed a freeze on subsidies and tax breaks for developers who burn the forests, they also initiated strong financial penalties. On August 30, 1989, the BBC reported that more than $10 million had been levied in fines against forest burners within just the past two weeks. And one particularly imaginative venture had been interrupted: an American television crew had arrived in the region to film the burning of the rain forests, in order to bring home to the American public just how extensive and terrible the situation really is. But finding that no burning was going on at the moment, the producer arranged with locals to burn purposely 4,000 square miles of forest so that he could get some spectacular footage. This venture was stopped at the last moment by someone in the Brazilian government (obviously someone who did not share the macabre sense of humor of the television producer).

But this effort will die out unless some meaningful legislation is actually passed in the United States, and there is strong opposition. The American automobile industry is not sitting back idly while we talk about taxes on gas-guzzling and CO_2-spewing Buicks and Chryslers, of lowered import duties on gas-efficient Hondas, or of upgraded legislation to make American cars more fuel-miserly. And the coal industry is not going to grin and apologize, and help us switch to alternative energies. These lobbies are the beasts under our beds, and when they go bump in the night we all get bounced around.

However, if the people can outlobby the lobbyists, and if the proposed governmental actions are made a permanent part of a global program, and if the industry follows the lead of Applied Energy Systems, and if the people gather together as the Students for the Children's Rainforest do, we'll be making a good beginning to finding the end of the problem.

But even all this is only a beginning, and in fact a beginning without an end—because we're not going to reforest areas the size of Australia every year. The only real solution is to find ways to stop burning massive amounts of fossil fuels.

That's not a revolutionary statement. We've always known we were going to have to do that eventually, because the fossil-fuel reserves are finite and nonrenewable in human lifetimes. But the greenhouse problem tells us we have to start doing it right away, and that means finding alternative sources of energy. Luckily, we know what they are and how to utilize them. Unluckily, they're not quite ready yet.

17

The Second Coming

And what rough beast, its hour come round at last,
Slouches towards Bethlehem to be born?

—W. B. Yeats, "The Second Coming"

The simplest alternative fuel raises no great problems, only a slight reorientation of our thinking. The burning of methane, surprisingly enough, provides less of a greenhouse threat than does the use of either coal or oil. It does, like these others, provide one molecule of carbon dioxide for every atom of carbon burnt, but it releases more energy per molecule than either of the others and so generates less carbon dioxide per unit of energy used. We have to be careful not to let the methane leak into the air, but current measurements indicate this isn't a problem and increased use shouldn't make it so.

Our methane reserves are not as extensive as our buried coal but are more plentiful than oil, and we could easily make more use of this "natural gas" than we do at present. It's not convenient as a transportation fuel, as explained earlier, but for such uses as home and hot-water heating it's excellent. The savings in atmospheric greenhouse gas levels will be noticeable but minor, not sufficient to halt the problem but—since we have the methane and since it's cleaner than either coal or oil—we should certainly use it in preference to the others.

The ultimate answer to the problem will be the use of fusion and/or solar power, but neither of these is ready yet. We'll discuss them in the next chapter. The immediate answer to the problem of destruction of our atmosphere is one that the environmentalists have been fighting for many years. It's time for the second coming of nuclear energy.

* * *

After the Second World War the world was promised a New Age based on the miracles of atomic fission, private airplanes, and television. Homes would be heated by nuclear power, people would be informed and entertained by television, and in every garage would be a small airplane or helicopter to take us wherever we wanted without the problems of crowded highways or polluted air.

It hasn't quite worked out that way. The little planes simply bounced around too much in the air to provide the image of safety to people who want to be able to stop and get out at will, and the industry never reached the mass-production level which would have made economical the abundance of private airports that would in turn have made the planes convenient and cheap enough for enough people to buy. Television has certainly come to dominate our entertainment lives, but when one considers the brilliance of the human mind in inventing such a system and contrasts it with the dull lack of imagination in programming it, one begins to wonder whether the human race *should* survive.

As for nuclear energy: Giant grids were envisioned, with cheap and virtually unlimited nuclear power sent all over the land. No longer would there be vast rural areas without light, heat, and plumbing; we could even run heating grids under highways to melt snow and ice as they formed. No longer would there be tens of thousands of deaths every year from coal-mining accidents or emphysema and lung cancers from the pollution effects of coal burning, nor would our cities continue to be enveloped by killer smogs like the December 5–8, 1952, London disaster, which resulted in more than three thousand deaths. The cover of *Life* magazine showed the miraculous new "smokestacks" at the first atomic plants, from which no smoke at all issued; they became a symbol of the cleanliness and safety of the new miracle.

And then, slowly at first, the new nasty words began to enter our vocabulary. Radwaste. Contamination. Meltdown. Strontium-90 and Iodine-131. Plutonium. Three Mile Island. Chernobyl . . .

The term *atomic energy* is somewhat misleading, since all chemically produced energy is ultimately atomic in nature: the burning of fossil fuels yields energy when an atom of carbon reacts with oxygen atoms. What we call "atomic" energy is generated by the fission of uranium and plutonium nuclei, and so *fission* or *nuclear* energy is the preferred term. (Nuclear energy can also refer to the fusion process, but since we won't be discussing that till the following chapter we'll use the two terms synonymously until then.) The process depends on the rather accidental abundance of the rare isotope uranium-235.

I use the word *accidental* because uranium in general and that isotope specifically are one of those atoms that just happen to exist. If carbon or oxygen or any one of a half-dozen other elements did not exist with the properties they have, *we* would not exist: they are a necessary ingredient of our planet and our life. Hydrogen is necessary to the energy process that makes the sun shine. But uranium, for the first twenty billion years of the universe's existence, was of no earthly or celestial use whatsoever. It is a by-product of the r-process of nucleosynthesis in stars, contributing nothing more to the stars than ash does to a wood-burning fire. Furthermore, it is an unstable atom and will ultimately disappear. All atoms with a mass greater than 209 are radioactive, unstable, and not long for this world.

"Not long" is a relative phrase, and it accounts for uranium's existence on this particular world at this particular time. The uranium we have on earth was created—as were all other elements except for hydrogen—in the interiors of stars that existed billions of years before the earth was created. This means that the production process stopped some time previous to four and half billion years ago, when the earth was born. Uranium-235 has a half-life of 0.7 billion years: Every 0.7 billion years half the stock of existing U-235 disappears by radioactive decay, ultimately becoming an isotope of lead. After four and a half billion years there is only one percent of the original number of atoms left. The process will continue until someday there will be no U-235 left on earth (ignoring the philosophical conundrum that you can never remove the last drop of water in a bucket by continually removing half of the bucket's contents). That's what I mean by uranium being a happenstance element. If its half-life happened to be ten times shorter, there wouldn't be enough uranium on earth today for fission to have been discovered or for it to be a viable commercial and military process. But God is good, and while U-235 has almost entirely disappeared, there is still enough of it around for us to use.

The more abundant uranium isotope is U-238, and there is more of it today simply because it has a longer half-life and so less of it has disappeared. Energetically, the important difference between them is that U-235 is fissionable with slow neutrons and U-238 is not. The fission process (discovered by Otto Hahn and Fritz Strassmann in Hitler's Germany during the late 1930s, and first understood and explained by their Austrian co-worker Lise Meitner as she was fleeing arrest by the Gestapo) consists of a neutron entering the U-235 nucleus and jiggling it enough to upset its equilibrium so that it breaks apart into two smaller nuclei. The two smaller nuclei have a total mass less than that of the original U-235, and so mass disappears in the process; this mass appears again as energy according to $E = mc^2$. The c-squared term is the square of the speed of light, an exceedingly large number, so that when it is multiplied by even a small amount of mass it results in a very large amount of energy being liberated.

The greatest—though by no means the only obstacle—to the utilization of nuclear energy today is fear. Some of this fear is well justified, but much of it is not. Because the fission process which takes place in a nuclear reactor is similar to that which takes place in a nuclear bomb, many people are afraid that the former could accidentally become the latter: that if a technician pushes the wrong button or if some salient, complex, mysterious piece of machinery fails, a commercial reactor could erupt in a fireball and vaporize a nearby city.

The fear arises because of our everyday knowledge of the world: we know that, except for scale, a chemical fire is the same thing whether it occurs in a burning cigarette or in a raging hotel fire, and the cigarette can start the hotel blazing if it is carelessly dropped. But a nuclear explosion is different in kind from a controlled chain reaction and cannot be caused by any accident in a reactor—even if it should go totally and helplessly out of control. This is because of a concept badly named *critical mass,* which follows directly from the logistics of chain reactions.

An illustrative analogy for a chain reaction is a table covered with mousetraps with their springs stretched into firing position, each baited with two Ping-Pong balls. If one more Ping-Pong ball is dropped into the midst of this array it will hit a trap, springing it; the two Ping-Pong balls originally on the trap will be thrown into the air. If these hit two other traps, four balls will be released to strike four other traps, and the number of balls flying around the table increases geometrically: within a few seconds all the traps will have been sprung, releasing their Ping-Pong balls and their stored energy. This is an explosive mousetrap chain reaction.

Suppose the table was a small one, with room for only two such traps. If the initial Ping-Pong ball was dropped on one of these traps, it would fling up its two balls—but these would be flung far beyond the adjoining trap and would fall uselessly to the floor without initiating any further reaction.

Suppose the table was large enough to hold a half dozen traps. The same result would probably happen. If by chance one of the two Ping-Pong balls went nearly straight up, so that it fell nearly straight down again and hit one of the half-dozen traps, setting it off, it is likely that the further two balls thrown up would miss the table: the chain reaction would die out.

In order to initiate and complete a chain reaction, a large enough table—large compared to the range of the flung Ping-Pong balls—must be covered with traps. In a large piece of U-235, if one atom fissions it not only breaks into two smaller pieces but it also releases more than one additional neutron. These fission neutrons hit another U-235 nucleus and fission it, causing the release of still more neutrons, and in a manner directly analogous to the mousetraps but millions of times more quickly, a chain reaction is initiated, a large fraction of the uranium atoms fission, and the released energy all comes out at once: we have a nuclear explosion.

In fact, such a large piece of U-235 cannot exist: it will spontaneously and immediately explode if any stray neutron enters it to begin the chain reaction. (There are always a few stray neutrons around, either from U-238 impurities, which spontaneously generate neutrons, or from cosmic-ray sources.)

But suppose the piece of U-235 is smaller. Then the fission neutrons may travel right through it and escape from it before they happen to hit another nucleus. Remember, the uranium atoms like all atoms are nearly all empty space: each consists of a central nucleus surrounded by a distant cloud of electrons. If the atoms were drawn to scale, and the nucleus were drawn as a circle one inch in radius, the outer edge of the electron cloud would be about three football fields away; the next closest nucleus would be six football fields away. And a single neutron would be the size of a large dot on the paper.

So think of this dot being flung out of the inch-wide nucleus. It goes hurtling through empty space, with another nucleus several hundred yards away. What is the probability that it will bump into that nucleus? Obviously very small. So a very great number of such surrounding nuclei must be in place for the neutron to have a reasonable chance of hitting any one and causing another fission. On an atomic scale, the mousetrap-laden table must be very big indeed.

The number of U-235 atoms necessary to sustain a chain reaction is known as the *critical mass.* It's a poor choice for a name, because it actually refers to a critical *density.* If the same number of mousetraps, for example, are spread out on an immense table so that the Ping-Pong balls are likely to fall in between them instead of on them, the chain reaction will not be sustained. In a similar manner, if a critical mass of U-235 is separated into many—or even two—components, it won't sustain the chain reaction.

This is how nuclear fission bombs are built. One calculates the size of a sphere of U-235 with a critical mass. (It comes to about six inches in diameter.) Then one fabricates two *half*-spheres. Neither of these has a critical mass. They can be safely handled, and each can be loaded into the opposite end of a single long tube, their flat bottoms facing each other. Behind them, at each end of the tube, is a compressed spring. To set off the nuclear explosion the springs are simultaneously released, the two half-spheres are pushed together into one whole sphere, *which now has a critical mass,* and bang!

In actual practice, of course, it's more complicated. As long as the two half-spheres are kept far apart in the tube, any neutrons from U-238 impurities may induce a fission reaction, but the resulting fission neutrons are lost and the reaction dies out. When the springs are released and the two half-spheres come flying closer together, some of the neutrons escaping from one half-sphere will hit the other and cause more fissions, and the resulting number of neutrons flying around in and between the two will increase, rising exponentially as they get closer together. Along with increased numbers of fission neutrons is an increasing release of fission energy, and before the two halves can come fully together there will be so much energy released that it will push them apart again, and the chain reaction will die down: instead of a nuclear explosion the result will be a nuclear fizzle. Enough energy and neutrons will be liberated to kill any people standing around watching, but nothing else will happen.

In order to make a nuclear explosion, the critical mass must come together within a fraction of a millionth of a second, before the buildup of fission energy can push it apart. The only "spring" strong enough to fling the pieces together so quickly is an explosive device, and such an explosive device will just blow the whole thing apart instead of together unless it is very carefully designed and machined and put together. A nuclear bomb is an incredibly complicated and precise piece of machinery; the chance of enough U-235 coming together in a nuclear reactor accidentally, in the exact configuration needed and quickly enough to explode, is about the same as the chance of a new baby being created by simply dumping the necessary amounts of carbon, hydrogen, oxygen and phosphorus atoms together in a box. It's just not the way babies are made.

A nuclear reactor, though basically the same thing as a nuclear bomb, is as different from it as Marilyn Monroe was from Shirley Temple. Visualize a swimming pool with a large number of vertical rods hanging in it. Half these are uranium fuel rods and half are control rods made of other elements, such as boron or cadmium, which will absorb neutrons with high efficiency and without fissioning. If it were not for the control rods, the uranium would constitute a critical mass; but enough of the neutrons emitted by uranium are absorbed by the control rods to break and stop the chain reaction.

The top of each control rod is attached to a wire leading up to a pulley arrangement, and now one by one each rod is lifted out of the "pile." As they are removed from the pile, the number of neutrons bouncing around among the uranium rods increases. As the control rods are slowly removed the pile approaches criticality as slowly and carefully as one wishes. Eventually it reaches the point where the neutrons are producing enough fissions to generate enough further neutrons to keep the chain reaction going. This is a nuclear reactor, and the energy generated by the fissioning uranium nuclei can be mechanically transferred to water, heating it to steam, which can then turn a turbine and generate electricity. If the reactor should reach criticality too quickly and accidentally flop over the edge, the result would be a nuclear fizzle rather than an "atomic bomb": the fuel rods would be pushed apart by the energy release and the reactor would quickly drop below criticality. We'll discuss this sort of thing on page 171. (There are many different variations on this basic theme, but they don't change any of the arguments that follow.)

There are just a few more pertinent points. The water in the swimming pool plays three important functions. It cools down the fuel rods so their internal energy doesn't cause them to melt, and it absorbs the radiation they emit so that people standing nearby are not injured. Most important, it slows down the neutrons from about 30 percent the speed of light to something less than the speed of sound. Enrico Fermi won the Nobel Prize in 1938 for demonstrating that such slow neutrons are much more effective in causing fission than the fast ones. The reason is simply that the high-speed neutrons zip through a nucleus so fast that they are gone before the nucleus even knows it's been hit; the nucleus returns to its original equilibrium state without fissioning. A slow neutron comes in and bounces and rattles around among the nucleus's neutrons and protons, causing them to break apart, or fission.

The slowing-down process is accomplished by bouncing the neutrons off other atoms, and the lighter the atom the more effective it is at slowing the neutrons. The hydrogen in water molecules, or the heavy hydrogen in heavy water, are the most effective for doing this, but graphite (carbon) can also be used.

A fission bomb cannot utilize the increased efficiency of slow neutrons (often called thermal neutrons because their velocity is that imparted by the heat of their surroundings rather than by the original energy of the fission event) because the process of slowing them down takes too much time. The bomb must explode within less than a millionth of a second or it will fizzle, and it takes thousands of times longer than that to thermalize the neutrons. Because of the high efficiency of thermal neutrons in causing fission, nuclear reactors can be fueled with natural uranium, in which the U-235 content is only 0.7 percent. A nuclear bomb must use highly enriched uranium, with a U-235 content more than ten times greater. This is another reason why a reactor using natural uranium cannot possibly undergo a nuclear explosion under any conditions.

There is just one more reaction of importance to discuss before we turn to the dangers that certainly do exist in a nuclear reactor. Most of the uranium is U-238 rather than U-235. A thermal neutron hitting U-238 will not cause it to fission, but will be absorbed and will form the new nucleus U-239, which is radioactive (with a half-life of twenty-three minutes). It will decay to neptunium-239, which will in turn decay to plutonium-239 on a time scale of days. Pu-239 is also radioactive but has a half-life of twenty-four thousand years, so that instead of disappearing on a human time scale it continually builds up in the reactor.

The interesting thing about plutonium is that aside from being an extremely poisonous and radioactive substance, perhaps the deadliest element in existence, it is fissionable by thermal neutrons; in fact, it has a higher fission efficiency (*cross section* is the technical term) than does U-235. Therefore it makes a better reactor fuel and also a better bomb.

It's also *easier* to build a bomb with it. U-235 is an isotope of uranium; in order to build a uranium fission bomb, the U-235 must be separated from the dominant U-238 isotope. The separation of isotopes is an extremely difficult business. It was probably the most difficult part of the Manhattan Project; Niels Bohr, the world leader of nuclear physics in the 1930s, predicted that the bomb couldn't be built without "turning all of America into one huge isotope separation factory." As it turned out he was nearly right: an entire city—Oak Ridge, Tennessee—was built to do this.

Plutonium, on the other hand, is a different chemical element with different chemical properties. If you were to take a fuel rod of uranium that had been irradiated in a reactor long enough to build up substantial quantities of plutonium, you would find it relatively easy to separate the plutonium chemically from the uranium. It is in this sense that reactors are most closely related to bombs: the spent nuclear fuel rods are inevitably and unavoidably a source of plutonium, which could be used by even marginally scientific countries like Iraq, Iran, or Libya to build nuclear weapons. But before discussing the implications of this inconvenient fact, let's discuss the other problems that stand in the way of our utilization of nuclear energy.

Because of its shielding and cooling requirements, a nuclear-energy-producing machine is necessarily a large plant. We will never have nuclear-powered automobiles or lawnmowers or flashlights; basically, nuclear energy is a source of electricity, and must be considered in relation to other sources.

The major difference between it and the other sources currently in use is its safety. This is a remark that surprises most people who are afraid of radiation, but the simple fact is that a nuclear plant in routine operation throws out to the environment less radiation than does a coal-fired plant of similar capacity. Coal is a dirty mineral, which during its formation incorporates, from its surroundings inside the earth, uranium and other radioactive elements; when burnt, these radioactive impurities are spewed out with the dirty smoke from the factory. The following table gives an idea of the amount of radiation people get from varying sources. (The data for coal and nuclear plants applies to people living nearby.)

	SOURCE	DOSE*
NATURAL	Cosmic rays	3
	Building materials	3
	Impurities in the ground	5
	Impurities in water	3.5
	Impurities in food	1.5
	Potassium in our bodies	1
MANMADE	Medical X-rays	10
	Cosmic rays while flying	3
	Luminous watch dial	1.5
	Color television	0.5
	Coal plants	0.5
	Nuclear plants	0.1

Nuclear plants in routine operation emit such low quantities of radiation because of their inherent mode of operation. A coal plant's refuse appears as hot smoke and *must* be vented to the outside air; a nuclear plant's refuse appears as fission products *within the fuel rods* and is simply kept there. The cooling water and the heat-transfer water (or other medium) that reaches the outside world does not have to come into direct contact with any radioactivity, and only minute traces of radiation should reach the environment.

Should, that's the frightening word. In routine operation a reactor is the safest possible way to produce energy, but what happens when something goes wrong?

If the control rods were all suddenly removed, the reactor would instantly flash to criticality—but "instantly" is a misleading word;

*The dose is given in roentgens averaged over a person's entire lifetime, and is an average value for people living near sea level. At higher altitudes the cosmic-ray dose can be much larger. The roentgen unit is chosen, but any unit (rad, rem, etc.) will do, and will not change the relative doses in any way.

an "instant" is a vague unit of time. The suddenly increased flood of neutrons would induce a spate of fission reactions, releasing energy and more neutrons, but without any confining pressure on the fuel rods the released energy would push them apart long before—many millionths of a second before—a nuclear explosion could occur; the result would be a fizzle rather than a bang.

It would be serious enough. The reactor would be destroyed and the metal enclosing the fuel rods might rupture, releasing vast quantities of radioactivity. The fuel itself might melt; this would be the dreaded "meltdown." But all the damage and all the radioactivity would be confined to the containment vessel, a secure pressure-proof metal and concrete enclosure within which the entire body of the working reactor is contained. *If,* of course, the containment vessel worked as planned and tested.

Would it?

That's the problem. Actually, simultaneous removal of all control rods is not a serious possibility for a nuclear accident or a sabotage event, but there are other scenarios that would lead to much the same result. Loss of coolant during operation could do similar things: The fuel rods would overheat and melt, releasing their radioactivity and simultaneously heating to the vaporization point whatever cooling liquid remained, possibly resulting in an explosion that would scatter the radioactivity all over the place if the containment vessel didn't hold. Something like this is what happened at Chernobyl.

At twenty-three minutes past one o'clock in the morning of Saturday, April 26, 1986, deep in the core of unit 4 of the N. I. Lenin nuclear power plant near Chernobyl, in the Ukraine, a few of the fuel elements suddenly went into a "prompt criticality" mode, almost instantaneously generating a hundred times more energy than normal. It happened so quickly that the uranium didn't even have time to melt; instead it exploded.

This was not a nuclear explosion, but an explosion triggered by the sudden generation of nuclear heat. The distinction is not a matter of semantics or nit-picking; the energy release was thousands of times less than a nuclear detonation would have been.

Even so, it was enough to do a lot of damage. It turned the cooling water to steam, and this generated a second explosion, one large enough to blow the top off the reactor and throw the intensely radioactive fuel elements out into the open air. Many of them were flung far enough to come down on other buildings, contaminating them instantly and setting them afire; the radius of damage was close to a kilometer.

There was a third explosion when the remaining coolant water and the steam reacted with the graphite that was used (instead of water) to slow down the neutrons, generating hydrogen and carbon monoxide gas. Hydrogen is explosively flammable when mixed with oxygen, and since the entire reactor was open to the air (especially with its top blown off by the previous explosions) it promptly exploded.

The cloud of radioactive gas was blown northwestward, passing over Scandinavia and then into Western Europe; traces have since been detected all over the world. Over a hundred thousand Soviet citizens were evacuated from the surrounding territory, many of them not in time: hundreds were hospitalized for acute radiation sickness; more than thirty of them died, and it is likely that hundreds more will eventually perish. Millions of gallons of water used to fight the fires were contaminated with radioactivity, and when this water ran off into nearby streams and rivers the entire water supply of the region was contaminated. As far away as Western Europe meat animals were slaughtered and buried, vegetables were harvested and thrown away, and people huddled indoors, afraid to breathe the air.

It was horrible. It was also stupid and unnecessary. There was no good reason for all of it to happen, but several bad reasons sufficed. The reactor was designed with a built-in danger. The type, known in Russia as RBMK, is unique in the world among commercial reactors, with an innate instability particularly dangerous at low power, owing to the way it uses both graphite and water. The water serves to transfer heat and to absorb some of the neutrons, but as it turns to steam in normal use some of its absorptive powers decrease. In other water-cooled reactors the water serves also to slow down the neutrons, and this capability will also decrease as the water boils; but at Chernobyl the moderating is done by graphite, which does not boil away. So as the water boils the neutron flux increases, and the slowing-down process remains the same. This increasing fission capability is balanced by a negative temperature gradient at high operating power, and by moving the control rods further into the pile when necessary. But at low powers the temperature gradient is insufficient to have much effect, and the control-rod movement takes a second or two to accomplish. So if the reactor is run at high power *and* any rapid changes in power level are limited to one percent (so the operators can compensate by adjusting the control rods), the reactor is safe. If either of these conditions is violated, the reactor is inherently *not* safe.

Early in the morning of April 26 the reactor was being run at only 25 percent power, in violation of regulations. At 01:23 A.M. the operators realized that the boiling of water was giving a rapid change in the power level of several times one percent. By the time they reached over and pushed the scram button to shut the reactor down, it had blown itself up.

That was problem number one. The second problem was that the reactor had been built without a containment dome. The Soviet engineers (or more probably their administrative and political bosses) thought that the thousand-ton lid was sufficient cover, but it was blown over on its side by the pressure of the erupting steam, and the hot radioactive fuel elements were scattered over the countryside.

Several years before, the president of the Soviet Academy of Sciences, writing in *Pravda* about the Three Mile Island accident and reassuring everyone of the safety of Soviet reactors, had said: "This accident [Three Mile Island] can happen only in a capitalistic society where they put profits ahead of safety."

Why did Chernobyl happen? The double design flaw, one involving positive feedback to reactivity from vaporization of water and the other being the elimination of a containment vessel, were due solely to economics. There was no valid scientific or engineering reason for either of them; the former gives a slight gain in energy per ruble, the second gives a saving in capital outlay.

And so it goes.

Could it happen here? We're told not. But we have a hard time believing what we're told about nuclear safety, and with good reason. Originally both nuclear development and licensing in this country were under the federal control of the Atomic Energy Commission. While this must have seemed a good idea at the time, it resulted in one group of people encouraging research and construction of facilities, and the same people deciding whether what came out of it was good or not. As Edward Teller is fond of pointing out, such a dual role used to be reserved only for God (Genesis 1:1–31). More recently these two responsibilities have been divided between the Nuclear Regulatory Commission and the Department of Energy, but—as with God—problems of belief and faith remain.

In 1979, after the Three Mile Island "incident," the nuclear industry not only won the yearly Doublespeak Award of the National Council of Teachers of English, but a new word was invented in their honor: "Nukespeak." Describing what had happened at the Pennsylvania power plant, they used terms like "normal aberration" and "abnormal evolution." The word "explosion" was never heard; it was replaced by "energetic disassembly." There was no fire, just a "rapid oxidation," and no plutonium contamination of the reactor vessel at all—although Pu did admittedly "take up residence" there.

It's no wonder we're worried about nuclear plants. Radioactivity is invisible, odorless, and deadly; and the government agencies that defend us against it are all too often just as invisible and just as deadly, if not odorless. In 1977 an AEC report forecast 3,400 deaths and 43,000 injuries from a reactor accident; it was suppressed until 1985 "to avoid great difficulties in obtaining public acceptance of nuclear energy." And things haven't gotten much better. In 1989 the Institute for Energy and Environmental Research obtained previously classified documents by invoking the Freedom of Information Act, a process the agencies purposely make difficult in order to discourage just such organizations. The institute discovered that a nuclear plant which produces enriched uranium for nuclear weapons, located near Fernald, Ohio, had discharged six times as much radioactive dust into the surrounding environment as the Department of Energy had ever admitted.

The plant has scrubbers that remove uranium from the exhaust fumes. In preliminary tests the scrubbers removed 83 percent of the uranium, and so the operators measured the amount that they removed and were thus able to calculate how much was being released to the outside world. But the newly declassified documents reveal that the figure of 83 percent was not constant; sometimes the scrubbers failed completely. Furthermore, the DOE and the plant operators knew this. Several internal memoranda acknowledge the problem, but these had never before been made public, had never been released by DOE or the operators. Measurements have now been made on the soil around the plant, and confirm that it is loaded with much more radioactivity than should be there.

The problem is that the nuclear plants operated for the military weapons program can be run under the cloak of secrecy, and safety flaws that would not be tolerated in a civil reactor are accepted and hidden. For example, although we have no civilian reactor with as dangerous a design as the Soviet RBMK, the Department of Energy has been operating a similarly graphite-moderated, water-cooled reactor at Hanford, Washington, to produce plutonium for bombs. It wasn't until the word got out about the similarity between it and the Chernobyl design, with a consequent public outcry, that the DOE finally shut it down in December 1988, "for safety improvements."

But despite all this official secrecy and these outright lies, the fact still remains that the only serious commercial accident in America, at the Three Mile Island plant near Harrisburg, Pennsylvania, on March 28, 1979—in which the operators made a series of six serious errors within fifteen minutes, culminating in the loss of reactor coolant—is just about the worst that one could even imagine happening to an American nuclear power plant, and it didn't kill a single person.

The fuel elements overheated, but neither melted nor burst. Some radioactivity escaped into the neighboring environment, but the levels experienced by people living nearby were less than they would get by spending a month in the mountains. When an airplane crashes or a coal mine caves in, when a dam breaks or a killer smog hits, there are death lists in the newspapers, sometimes hundreds of names long. *There were no death lists at Three Mile Island.*

True, the long-term effects aren't in yet. There may have been sufficient release of radioactivity to cause a few cancers or birth defects in the exposed population, but that isn't much. And if that sounds callous, consider the alternatives. Coal-powered plants *routinely* cause *thousands* of deaths *every year,* not only from their radioactive emissions but from the coal dust itself. Remember black lung disease, emphysema, and pneumonia? Remember coal-mining cave-ins? And remember the London smog of 1952 and its three to four thousand deaths?

No, Three Mile Island wasn't bad at all, even if the long-term effects are much, much worse than all the present measurements indicate they will be. And it's the *only* incident of its kind ever to happen in the United States, and what we learned from it will help ensure that it will never happen again.

The truth is that nuclear plants are the safest form of large-scale energy production ever invented by mankind. And yes, that includes the "safe" forms of energy such as natural gas and dams. You probably don't remember March 18, 1937, when the gas line feeding a New London, Texas, school leaked and ignited, and the explosion killed 410 children and teachers; or the Johnstown, Ohio, flood of May 31, 1889, when the South Fork Dam broke and killed more than two thousand people. But do you remember the Canyon Lake, South Dakota, dam, which broke on June 9, 1972, and left several hundred dead? Or the Macchu Dam in Gujarat, India, which left thousands dead when it collapsed in 1979? By contrast, the expected number of fatalities if we were to have one hundred reactors operating at full power in the United States is four per year. In the past thirty years there has been not one single death or cancer attributed to commercial nuclear power among the several thousand workers in the industry. Yes, there was the terrible accident at Chernobyl, but such accidents happen with or without nuclear plants. Just three years later, late in the night of June 3, 1989, Soviet workers pumping natural gas into a pipeline in the Ural Mountains ignored warning signs which should have indicated to them that the pipeline was ruptured. An hour after midnight the gas, which had leaked out and infiltrated a mountain valley, erupted just as two passenger trains were passing by. Hundreds of people died in the explosion, many hundreds more are in hospital; the death list is longer than the one at Chernobyl.

There are four sides to nuclear-reactor safety. The first is the media presentation and public perception of the dangers, which are disproportionately flung about. For example, *Time* devoted just one paragraph to the Macchu Dam disaster, which killed thousands; contrast that with its coverage of Chernobyl, which killed hundreds, or Three Mile Island, which killed no one.

The remaining three sides have to do with the actual dangers of nuclear power. One side concerns the routine aspects of running the reactor: containment of radioactivity, environmental effects of thermal runoff, and the disposal of radioactive waste. The problems associated with all of these have been solved at least in principle, and are solvable for each particular case by further applied research. Radwaste, for example, is a serious economic problem because it has to be disposed of carefully, but containment vessels have been developed that have no serious risk of leakage, and geological sites such as salt deposits have been found where the waste will be safe for thousands of years, long enough for it to decay to natural levels.

Are these disposal schemes *absolutely* safe? No. What is? Are coal plants safe? Are dams safe? Is the greenhouse effect nothing to worry about? If you want absolute safety, lead a good life, die, and go to heaven. In *this* world, nothing is guaranteed. But nuclear power is a better bet than anything else we have.

The third aspect of nuclear safety is the possibility of a vast accident like Chernobyl. The frequency with which these will occur is impossible to calculate precisely, because of the nature of improbable events. And these events are *exceedingly* improbable. Impossible, no; but the sum total of deaths resulting from them will clearly be way below the total of deaths resulting from any alternative form of energy production. So the answer must be to cut down on our energy needs as much as we can, and then get on with nuclear energy. The only thing we really have to worry about—we have to be *careful* about everything concerned with nuclear energy, but the only thing we have to *worry* about—is the fourth side: plutonium.

The major problem facing the United States when we tried to introduce the world to the benefits of nuclear energy with the 1953 Atoms for Peace program was the relative ease with which plutonium can be made into bombs. The solution proposed then, and still the best one around, is for any nation receiving nuclear materials or help of any kind to agree not to attempt to separate plutonium from its used uranium fuel rods, and to open inspection of all its nuclear facilities to guarantee that commitment. The Non-Proliferation Treaty of 1968 formalized this convention; in it all nuclear nations agreed to abide by its provisions when supplying aid to newly developing countries. But France, among others, wouldn't sign the treaty and has since gone its own way. With French help Iraq not only built a reactor but was well on its way to building a bomb until Israel attacked and destroyed the reactor before it went critical.

The process of extracting plutonium from irradiated fuel rods is simple, in terms of the Manhattan Project, but it is certainly not a bathtub operation. Huge reprocessing plants are needed, so the process can't be hidden. The problem is that reprocessing is also used for non-weapons purposes, chief among which is the *breeder* reactor.

This is potentially the most valuable type of reactor, and definitely the most dangerous. It takes advantage of the plutonium buildup to generate new fuel. When nuclear power was first realized, in the late 1940s and 1950s, it looked as if it would be a nearly limitless operation. But uranium exploration during the following years indicated that there were not many economically useful deposits, and the forecast was that if the world began to run largely on fission power we'd run out of uranium within fifty or a hundred years.

That is because the normal reactor fuel consists of U-235, which constitutes only 0.7 percent of the total uranium. If we could find some way to use the remaining 99.3 percent we could extend the nuclear energy lifetime from fifty years to more than five thousand years.

Plutonium will do it, and the method is conceptually simple. During operation of a normal reactor, some of the neutrons are captured by U-238 to form plutonium. Calculations show that an attainable increase in the capture rate will produce more plutonium during the reactor's lifetime than the original amount of U-235 fuel. Theoretically, nearly all the U-238 can be converted to plutonium under the proper irradiation conditions.

The "proper irradiation conditions" include the use of fast neutrons rather than thermal ones. But this lowers the fission efficiency of U-235, so fuel of enriched uranium (enriched in its percentage of U-235) must be used.

And so we have the *fast breeder* reactor, which can generate all the electricity the world will need for thousands of years, and which is doubly deadly. First, since it uses fast neutrons it starts with an enriched fuel—and an enriched uranium fuel is suitable for making bombs, whereas a natural uranium fuel is not. A nation could simply order a fast breeder from France or any other nation willing to supply it without safeguards, and use all or part of the fuel to make bombs instead.

Second, the fast breeder quickly builds up vast amounts of plutonium, which can be extracted and made into bombs even more easily than the enriched uranium.

Because of these problems many voices have been raised against the fast breeder, and in fact the United States has killed its own development program. But despite the problems—or perhaps because of them?—other nations are plunging full speed ahead. They are simply following our lead, without changing their minds as we did. In 1967 our AEC made fast breeders its number-one priority, spending $250 million per year, an amount equal to 40 percent of its energy budget. We wanted four hundred plants in operation by the year 2000. By 1970 we had the Barnwell Reprocessing plant in operation at Savannah River, Georgia, with the capacity to handle the radwaste from sixty nuclear reactors a year, recovering fifteen tons of plutonium every year.

In that same year Germany, Italy, Argentina, India, Japan, and Pakistan began their own reprocessing operations. None of these had signed the Non-Proliferation Treaty. In 1973 a study indicated that the uncertainty in these reprocessing plants amounted to about one percent. The uncertainty refers to the error in the amount of plutonium extracted. Losses or thefts within that margin of error cannot be detected, and an error of one percent means that a typical reprocessing plant could "lose" 150 kilograms of plutonium a year—enough to make 15 bombs.

In 1974 India joined the countries possessing the nuclear bomb. It was a plutonium bomb, the plutonium coming from its reprocessing plant designed for the reactor that had been supplied by the United States and Canada.

By 1977 the United States had decided that because of the problems of weapons proliferation, reprocessing should be canceled. Uranium reserves, we announced, were larger than expected and future electricity demands were anticipated to be lower than expected. There should be sufficient uranium in the world to take us well into the next century, without breeding plutonium.

The proliferation problem was directly traceable to reprocessing. Six of the seven countries that had tested bombs by then had used reprocessing to get the fissile material. But our appeal to international sensibilities fell on deaf ears. France and Germany announced they were planning to export reprocessing plants to Pakistan and Brazil. (Pakistan was urgently searching for help in obtaining "peaceful" uses of nuclear energy, particularly since India had built a bomb a few years previously.)

In 1980 an International Fuel Cycle Evaluation meeting was held in Vienna at U.S. instigation. Sixty-six nations attended and listened to our arguments to scrap the breeders because of the Pu proliferation problem; but they didn't agree to do it. The have-nots wanted nuclear power without restrictions, and the other nuclear nations wanted to catch up to us and hopefully even pass us in the nuclear game. We argued that there was ample uranium for the whole world for the next century; the other nations disagreed. (They also weren't particularly overjoyed that much of the world uranium reserves we were referring to was located in the United States.) They countered our arguments by pointing out the safeguards imposed by the International Atomic Energy Agency, and then they all went out for drinks and had a good laugh over that.

The IAEA safeguards are practically nonexistent. The home country must be notified of inspections ahead of time, and if they don't manage to get things cleared away and hidden before the inspectors show up, they can simply say they're not ready and refuse to admit anyone. The IAEA can't do anything about it except complain, and as a last resort bring the matter up to the United Nations Security Council—the punitive powers of which are well known to one and all. There is really nothing the IAEA can do to find a serious problem the host country wants to hide, or—if they should find it—to do anything about it except expose the violation. And of course any country that wants to can always resign its Non-Proliferation membership and join France, China, South Africa, and all the others.

We in the United States can't simply look at the international situation and blame everyone else. We play this game as well as anybody, as is illustrated by an old Indian fable, which goes like this: Since 1963 the United States has been committed to supply the fuel for India's commercial reactor at Tarapur, which was built with U.S. aid. But in 1974 India exploded a bomb (using plutonium obtained from their own reprocessing plant that was *not* part of the Tarapur complex); they had joined the nuclear-weapons club without permission, which violated the Non-Proliferation Treaty. And that put us into an embarrassing position. We didn't want to cut off our nuclear aid to them for diplomatic reasons; India is a strategic country in the Far East, a big-stakes player in the China-Russia game, and we didn't want to walk out on our "obligations" and give Russia or China unopposed entry. But according to the terms of the Nuclear Non-Proliferation Act passed by Congress in 1978, we may not give any nuclear fuel or equipment whatsoever to a nation that makes a bomb.

India saw this as an attempt to renege on our 1963 agreement, which according to their point of view was still in force; what they might or might not have done at some place other than Tarapur didn't affect the Tarapur contract, they argued. They would not allow inspection of any of their nuclear facilities except Tarapur, and if we would not supply fuel for Tarapur they would cut themselves off entirely from the United States. Congress saw it differently: any further nuclear aid to India was prohibited by law. The Reagan Administration stood between the two, looking anxiously from one to the other while Russia happily tiptoed in selling India a squadron of its most advanced fighters, the Mig-29, and loaning them a fully equipped nuclear submarine.

The administration finally came up with a diplomatic solution: they got France to supply the fuel. France was happy to have the business, and didn't care what India did with any plutonium that might result. India was happy to get the fuel, no matter who provided it, and in return for our working out the deal with France they agreed to maintain the previous close diplomatic ties with the United States, and to keep Tarapur open for inspection—though their other nuclear facilities, particularly the one where the spent fuel was reprocessed and plutonium extracted, would be kept secret from prying eyes. The U.S. Administration was happy to keep India in the Western camp. And the U.S. Congress was happy to accept the *pro forma* enforcement of their law and to have the weight of the problem shifted from their overburdened shoulders.

And the rest of us: are we happy?

Clearly plutonium reprocessing and the fast breeder are dangers to humanity. Do we need them? World uranium resources are known to top 6 million tons, with 24 million more still speculative. (There are 4 billion tons of uranium dissolved in the oceans, but we don't know how to get it out economically.) A standard commercial reactor, with an output level of 1,000 megawatts, uses 150 tons per year, so the world's known uranium reserves can fuel 30,000 gigawattyears of operation. There are currently 400 gigawatts of energy being produced by reactors (this includes the contribution from reactors under construction), so we have enough uranium for nearly one hundred years of operation at our current levels. But if we are to increase our production of nuclear electricity in order to cut down on fossil-fuel use, that hundred years will dwindle fast.

Are breeders the answer? While we worry about the plutonium buildup, we should recognize that the negative environmental impact from breeders is less than from any other form of energy production, since all the uranium needed to operate them for hundreds of years has already been mined. And they can go longer without refueling than normal reactors, which means fewer reprocessing plants are needed; we might even be able to handle the world's requirements with only a few reprocessors built in secure areas and under continual guard. They involve a higher capital outlay than does a normal reactor, but over their entire lifetime they produce electricity more cheaply since the fuel cost is lower. Today they are directly competitive with fossil-fuel plants, and ahead of anything else if one allows them one arguable and one non-arguable point.

France's Super-Phénix breeder is the world's first full-scale commercial breeder reactor. It was designed to lead the way into the next generation of breeders, providing electricity at 15 cents per kilowatt hour, which is a bit cheaper than nonbreeders. (Electricity from plants using solar power costs 50 cents.) But its capital outlay has involved a bit more than $10 *billion,* a sum so large that the reactor was jointly financed by and is owned by France, Italy, England, Germany, Holland, and Belgium. (The plan being that these other countries would learn the technology as the plant was built, and would then go on to build their own versions.) The incredible initial cost is justified by the arguable point that the breeder will probably last twice as long as ordinary reactors, remaining operable and profitable for one hundred years instead of forty. This is arguable because obviously there are no old breeders in operation, so no one can be sure. But it looked like a good bet and, if true, the breeders, like dams, will provide cheap electricity once they are amortized.

The non-arguable point—non-arguable because no one cares to argue it in public, for obvious reasons—is that a hidden profit comes from the military value of the plutonium produced, which will make Europe independent of foreign (United States) resources forever.

This is neither a trivial nor an imaginary component of value. In the United States we have separated commercial and military applications as cleanly as we have separated Church and State (that is, not all that cleanly; see Sandra Day O'Connor's 1989 opinion that "we are a Christian nation"). Our weapons plutonium, however, is generated purely by military reactors, and these are currently in need of renovation at a suggested cost of many hundreds of millions of dollars. An alternative being considered by the Department of Energy is an entirely new plutonium-producing reactor, at a cost of $4 to $8 billion.

As it turned out, the Super-Phénix has had more problems than anticipated. The cost of electricity it has provided is actually double that of the nonbreeders France is also operating, and a small leak of coolant has developed. No radioactivity is escaping, but since the Phénix uses liquid sodium as a coolant, and since liquid sodium reacts explosively with either air or water, the leak must be fixed. But the reason no radioactivity is escaping is that the whole reactor is sealed up tightly, and for that same reason it's terribly difficult to get to and fix the leak. It will probably cost about 400 million francs to do the job.

All in all, the European community is not happy with the breeder. The follow-up plants originally to be built by England, Germany, and France have been put on hold, and the European Community is now in the midst of a two-year study to determine how and if they should proceed. As this book goes to press, England has just cut the guts out of her reactor research funds. Japan, on the other hand, is proceeding full speed. Her Monju breeder will go critical in 1992, leading the way to a planned 60 percent generation of all electrical needs by the year 2030. Her plutonium safeguards, incidentally, are the best in the world, with IAEA inspectors permanently installed and on duty twenty-four hours a day.

To my mind it is clear that if we lived in a world ruled only by science, or even by science and economics together, the breeder would be the way to go. But there is a political component to this wicked world, and in that component are wicked people. "Wicked" is a term not well liked by social relativists, who believe that good and evil are as relative in this world as motion is in Einstein's, and so I had better define my terms. By "wicked" I mean people serving a cause which shines so brightly the light blinds them to the existence of other people, allowing and even encouraging them to plant bombs on airplanes and school buses, in pubs and synagogues; people who flood their battlefields with poison gas, killing friend and foe alike; people who command their followers to murder a writer who may have slandered them; people who assassinate the enemies of their gods.

There are people like this in every country in this world, and in some of them they have reached positions of leadership and dictatorship. They include not only Colonel Qaddafi and Yasser Arafat, but also the President of the United States, who sent a fleet of F-111 fighters to kill Qaddafi and/or his family. (Qaddafi escaped, but one of his daughters was murdered.) I'm afraid of those people; I don't trust them. If a president as mild as Eisenhower seriously contemplated using the bomb in a first strike against the Soviet Union, I am limp with relief that somehow we got through the Reagan years unscathed.

I'm not happy that any of these people have the bomb. We can't turn back the clock and take nuclear weapons away from those who have them, but I don't want any others to get them. I don't want Iran or Libya to get the bomb; I don't want the PLO or the IRA to steal or be handed enough plutonium to build a bomb. And as the breeder proliferates, so inevitably will the flow of "lost" plutonium.

This is a problem that science can't solve. We have a tendency to be spoiled in this country: we demand and expect technical solutions to all our problems, such as curing the drug problem by putting state-of-the-art AWACS spy planes in the skies to interdict South American drug supplies. But technical solutions don't always work, particularly when the problem lies not in the stars but in ourselves, not in the laws of the physical universe but within our own human weaknesses.

Unfortunately, there is no shortage of people who will sell nuclear bomb materials. In 1968 a shipment of uranium "yellowcake" ore went missing on the high seas. No one knows what happened to it, but if I were a betting man I'd bet it's in Israel now, in the warheads of a few bombs. In the 1970s and 1980s a West German firm shipped nuclear fuel (probably including plutonium) to both India and Pakistan, as well as to Argentina and South Africa, and possibly Libya. The German government insisted that it hadn't happened, until American CIA information showed that it had. Taking advantage of the most lenient nuclear-related laws in the international community, the German firm contracted to import the nuclear materials from Norway, China, and other countries. But they never reached Germany. The shipments were rerouted to Switzerland, then to Dubai, and then simply disappeared in a maze of international paperwork. The West German government is currently conducting an investigation, but they insist that the entire operation is simply not covered by German law, so what can they do? It's just good business.

I really hate to say it, I hate to admit it, but we're not ready for either plutonium or the fast breeder reactor. Scientifically they're beautiful things, but *Realpolitik* cannot be denied. We have to look at our choices, from the beginning to the end, and decide what to do. The beginning choice is whether or not to keep on burning fossil fuels, and we have to decide we can't go on doing this at our currently increasing rate because of the greenhouse effect. The only alternative at the present time is increased use of nuclear power. It can be made safe enough, and we should get started again on it. This means constructing new nuclear plants, and it means paying full attention to safety requirements, and that costs money and time. But the time to do it has come. It's a rough beast, but it's time for it to get moving toward Bethlehem, time for it to come again.

The breeder, however, is just too rough, too dangerous for this world. As a technological solution to our energy needs, it's a beauty; but a technological fix is not sufficient. It would be nice to believe that all our problems can be cured by science, just as it would be nice to believe that we'll be taken care of by God, but I'm afraid one belief is as false as the other. So until we can come up with the permanent solution to our energy problems we should get started replanting trees, conserving energy, and building nuclear reactors.

And does a permanent solution exist? Yes, if we accept "permanent" to mean the entire foreseeable future. It not only exists, but it's almost within our reach. Almost, but not quite . . .

18

The Solar/Fusion Fix

The answer, my friends, is blowing in the wind . . .
—Bob Dylan

There is nothing permanent in this world, not even death and taxes, because ultimately the universe will come to an end—either by fire or ice. It is currently expanding and has been doing so since its creation some fifteen or twenty billion years ago. Eventually this expansion will either continue or stop. If it continues, every galaxy will recede from every other until they are all alone and isolated in empty space; eventually every star will use up its fuel and burn out, every planet will freeze, and the universe will be nothing but a frozen waste. If the expansion slows and stops, gravity will take hold and pull the galaxies together again until they all collide and explode in a universal fire.

So, one way or another, nothing in this universe is permanent. But all that is many billions of years away. If we consider the state of things as far as we can possibly see into the future, a "permanent" solution to our energy needs already exists, at least in principle. In fact, there are two solutions.

Solar energy is one, and in a sense we already use it since except for nuclear energy *all* energy on this earth is ultimately solar in origin. Fossil fuels are simply the end products of the harnessing of the sun's energy by living plants, which use it to grow their bodies at the expense of their surroundings. To take carbon dioxide from the air, as plants do, and rearrange it into organic carbon molecules necessitates the expenditure of energy, which the plants get from the sun by photosynthesis. This energy is stored within the molecules, and is released upon burning.

Even the energy of Niagara Falls and of windmills is ultimately solar. Without the sun, all the water on earth would have flowed to the lowest basins billions of years ago; but because of the sun, water is continually evaporating and raining down upon the continents, forming into lakes and rivers, and flowing back to the low-lying oceanic basins. Along the way it can generate energy for us, as at Niagara Falls. The winds, too, are generated by the sun as it unevenly heats the earth's surface, which in turn unevenly heats the atmosphere, causing pockets of rising warm air and falling cool air, which in turn drive the entire circulation we call winds.

But all these processes capture only the minutest part of the sun's energy, which flows down to us freely and continually. They illustrate both the promise and the problems. The promise lies in the fact that the sun pours out to us more energy than we need; most of this is wasted—as when it bounces off the earth and is lost to space—or is even harmful, as when it is absorbed by the earth and then reirradiated in the infrared to heat the atmosphere more than we want. The problems in utilizing this excess energy are all tied up with its dispersal: instead of being concentrated, the sun's energy is distributed over the entire surface of the globe and is nowhere hot enough for us to use directly and efficiently. If you put a glass of water out in the sunlight on a hot summer day, it will get warm; but it will never boil. If sunlight could boil water directly, we could use the steam to drive turbines and all our problems would be solved.

But it can't, and they aren't; and so we strive to harness just some of that everlasting sunlight in a useful way. In a sense, it's the same problem we have in our bodies: they can't use all the energy we take in by eating (in an affluent society). Over the eons, we have evolved a mechanism for storing the energy; we pile it up in our bodies as fat, to be used when food is scarce (or when we diet). If we could utilize that stored energy at will, we would not only be as thin as we like, but we could leap tall buildings at a single bound and run faster than a speeding locomotive: we could all be supermen and superwomen. But we can't, and we're not; all that stored energy just makes us fat.

It's the same problem with solar energy: it's there for us to use, if we could only figure out how to channel it usefully. There are many visionary ideas on how to do this. Freeman Dyson, whom we mentioned earlier in connection with nuclear winter, has suggested that if we want to find advanced alien civilizations on other planets around other stars, we shouldn't look at known stars at all, because in a really advanced civilization they would be hidden. Using the theoretical knowledge we already have, and speculating on advanced methods in engineering that might become available thousands or millions of years in the future, he suggests that we (or others) might be able to take our solar system apart and reassemble the planets into one hollow sphere surrounding the sun. This would capture *all* the sun's energy instead of letting most of it escape into space, and we (or they) would use the inside of this hollow sphere as our world. Such *Dyson spheres,* as they have come to be called, would close off the sun from the outside; if other, older civilizations have already done this elsewhere in the galaxy, we won't find them by looking at normal planets around visible stars.

Well, it's a lot of fun to think about ideas like that, but we're not going to be reassembling our solar system during the current administration or any time in the foreseeable future. We could possibly, however, turn the small fraction of sunlight that reaches the earth into electricity or hot water through direct conversion schemes. The simplest to visualize is a series of solar panels in the roofs of houses, which would let in sunlight to heat our household water; slightly more complicated are *photovoltaic* cells, which can use the sun's energy to drive chemical reactions to generate electricity.

Since the sun's energy is free, why not use these schemes? The answer is, they're just not yet economically or environmentally competitive with other energy sources. Or at least not on any large scale. Direct solar hot-water heating for houses is a reasonable process at least in areas of high levels of sunshine. In southern Florida in the 1930s there were fewer than ten thousand such solar heaters; in the 1940s, during the war years when oil supplies were hard to get, the number rose to about thirty thousand. After the war it declined again, and now is slowly rising.

The principle involved in solar hot-water systems is simple, and so are the problems. Basically the principle is the same as the one in our greenhouse problem: a glass window allows the sunlight in to warm the water directly, and the glass traps the infrared reirradiated from the water, so the water warms up as does our atmosphere under its carbon dioxide blanket.

The problems are just as simple. Glass costs too much, and so does the necessary strengthening of the foundation to enable the roof to bear safely the very large additional weight of all that water. To equip a new million-dollar home with solar panels doesn't represent too much of a cost increase, and it's chic, so some new homes in places like Miami and Tucson are being built with solar water heaters. But to put these into cheaper housing *does* represent a significant increase in base price, and that puts houses with solar heaters at a disadvantage compared to others more traditionally equipped, and in today's tight housing market that means disaster. So nobody is doing it.

The problem could be solved partially by using lightweight, cheap plastic substitutes for glass. But the substitutes so far invented have serious long-term problems: they break down under ultraviolet radiation and become opaque and/or brittle. And since their function is to be exposed to sunlight, there's no way they can be shielded from the ultraviolet component of the sun's radiation, and the problem will get worse if the ozone layer keeps getting depleted. And there's still no cheap way to build the roofs strong enough to bear the added plumbing weight. So solar hot-water heating is an idea whose time has not yet quite come.

There are many other variations on this theme, many other ways to trap the sun's energy and deliver it to us in a useful form. The underlying idea is always the same: if two things are heated unevenly, there will be a flow of heat from warmer to cooler. A flow of heat is a flow of energy, and can in principle be made to do useful work, if we are clever enough.

The Israelis have been working on the *solar pond* concept for several decades. The idea is to dig a hole, line it with a black substrate, then fill it with water. The sunlight will penetrate the water and heat the black absorbing bottom and sides, which will in turn heat the water. If you could get it to heat the water to boiling, the resulting steam could be harnessed and put to work driving an electricity-generating turbine. The problem is that the heating will take place at the bottom of the pond, since it's the black bottom that absorbs the sunlight and provides the heat. But then it's the bottom waters that will get the warmest; as they heat up they will expand

and rise to the surface, the cooler surface waters will sink, and this circulation will keep any part of the pond from getting warmer than any other and from heating up to the boiling point, since the heat is continually being brought up to the surface and from there lost to the atmosphere.

The Israelis have tried *graduated saline* ponds. In this scheme the bottom water is made very salty, so that it's heavier than the water layers above it. As it gets hot it expands, but because of the salt content it's still heavier than the waters above it and so it doesn't rise. If the pond is carefully engineered so that the salinity decreases toward the surface, all circulation can be cut off and the bottom water just sits there getting hotter and hotter. They've been able to generate temperatures nearly up to 100° C. at the bottom, but unfortunately salty water boils at a higher temperature than pure water, and so no steam is generated; consequently the energy efficiency is less than one percent, and it's not economically viable even in a sun-baked energy-poor country like Israel.

There are many other similar schemes currently being pursued. For more than a decade the United States has been trying to get energy out of the OTEC project: Ocean Thermal Energy Conversion. This scheme uses the natural warming of ocean surface waters, transferring the mild heating to a solution of ammonia. Ammonia-water solutions boil at very low temperatures, and the resulting "steam" then drives a turbine and is in turn condensed by cold deep-ocean waters.

In essence, this is a technique to utilize the slight but real temperature difference between surface and deep-ocean waters. In principle, it sounds great: no pollution, free energy, and everyone lives happily ever after. In reality, it doesn't quite work. Ten years ago a small-scale electricity-generating motor that ran on this principle was built by a combination of Swedish and American private firms and the state of Hawaii. Mounted on shipboard, it was able to generate enough electricity to keep itself running; it was thought that a full-scale plant could be operating with positive electricity generation within ten years. But those ten years have come and gone and no such plant exists. (And even if the problems are solved and the scheme runs economically, it's not quite true that it will be pollution free. The impact on oceanic circulation patterns and on marine ecosystems of redistributing the cool deep waters on a scale large enough for a commercial plant is not at all clear. There is also the long-term problem of biofouling; the ocean is an environment thriving with life, and life does tend to get in the way of progress.)

The main alternative to conversion of sunlight into heat and thence into electricity generation is the direct conversion of sunlight into electricity through *photovoltaic* cells. This is not a new idea; the first successful attempt was reported by Antoine César Becquerel in 1839 (he was the grandfather of Antoine Henri Becquerel, who discovered radioactivity in 1898). The idea is simple. The effect of sunlight on certain crystals is to knock electrons loose from their atoms. These electrons may then drift through the crystal. Connected to the first crystal is a second dissimilar one, with a different potential for freeing electrons. The difference creates an internal disequilibrium, which translates into an electron pressure, or voltage, that can drive a current through an external circuit.

By 1954 the Bell Telephone Laboratories were making photovoltaic cells with efficiencies of a few percent, and by the end of that decade satellites were being powered by them. But they remained too expensive for routine use here on earth. However, recent experimental advances, combined with a rising cost of alternative (fossil) fuels, have brought the photovoltaic technology to the brink of effective competition, at least according to those enthusiasts within the research and development branch of the industry. The Solar Energy Research Institute is one of the leading advocates; its director, H. M. Hubbard, argues that solar power can be the main contributor to the United States' electricity needs before the middle of the next century. Even more important, solar power should be ready to be used as the power source for those developing nations that are not already committed to national electrical grids based on fossil-fuel supplies. All this, of course, if optimistic predictions about continued advances in technology and continued decreases in the cost of photovoltaic cell expenses hold true.

And there are other ideas. In England they're working on a sort of artificial photosynthesis. Their aim is to use the sun's energy to split off hydrogen atoms from organic molecules such as alcohol or amino acids; since hydrogen is flammable it will make a useful fuel (though a dangerous one; remember the Hindenburg?).

The concept of solar energy is so attractive that people have been working on it for decades all over the world, but it's not easy, and it's not yet ready. The sunlight is free and eternal, but the materials needed to harness it into something useful are not. The materials needed to collect the sunlight and turn it into electricity have to be mined out of the earth and fabricated into semiconductors, and before a system can be thought of as energy producing it not only has to produce energy but it has to produce more energy at less cost than was used in making its component parts; and in this world it has to do it more economically than alternative processes.

It also has to do it with a reasonable degree of safety, and this is where the solar enthusiasts are most enthusiastic. "What," they ask, "could be safer than solar energy?" Surprisingly enough, the answer is: nuclear energy.

That's a shocker, isn't it? But ten years ago the Atomic Energy Control Board of Canada took a long, hard look at the inherent safety and dangers of various energy schemes; their object was to determine whether Canada should "go nuclear" or seek alternative methods of energy generation. When they added up all the risks to the general public, they found that solar energy was nearly *five hundred times* more dangerous than nuclear. (Coal and oil were nearly two thousand times more dangerous; natural gas was the safest of all, but that neglected the greenhouse effect.) The reason for the public risk attendant on solar energy lies in the mining and manufacture of its components, in each stage of which other energy sources are necessary. For example, to mine copper for tubing necessitates an expenditure of energy, and the heavy machinery needed almost certainly means that coal and oil will have to be burnt. Because solar energy is by nature so diffuse, greater amounts of work and energy are needed to harness and concentrate it, and this translates into greater risk to the public through pollution, greenhouse problems, mining and construction safety, et cetera and so forth.

So solar energy is not free and it's not devoid of danger and risk, but the promise it holds out is enticing, and the more energy and the more money we put into the necessary research, the sooner we'll make it work efficiently and safely. The decision of how much solar research to fund, however, is complicated by a competitor: thermonuclear fusion.

Nuclear fusion is in different ways both the opposite of and part of solar energy. Sunlight is free and obvious, shining down on us everlastingly; it doesn't take much of a brain to think of using it. Nuclear fusion, on the other hand, is an energy hidden deep within the atomic nucleus; it took scientific genius to find it, liberate it, and conceive of controlling it for our own purposes. And yet nuclear fusion is the energy process that drives the sun, and in that sense the sunlight that reaches us is fusion power *redux*.

Our first nuclear fusion bomb ("device") was exploded on Bikini Atoll on November 1, 1952. In it, as in the sun, hydrogen was fused into helium, and the mass difference was released as energy. The mass difference is greater between hydrogen and helium than between uranium and its fission products, and so the energy released is greater—a thousand times greater, changing the face of warfare forever, making realistically possible for the first time the age-old fear of total extinction of the human race.

It also made realistic the dream of unlimited and safe energy. The fusion reaction is so difficult to ignite and keep ignited that if anything should go wrong the reaction would simply stop. With a thousand times more energy release than nuclear fission, with a great reduction in deadly radioactive waste, with no production of weapons material as a by-product, and finally with plentiful hydrogen as a fuel instead of scarce uranium, fusion power seemed at first to have no drawbacks.

Indeed, the government thought so. Awestruck at the energy blazing out of the fireball, they descended on the fusion scientists and told them how wonderful the bomb was and asked them when they could have a controlled fusion reactor for peaceful uses. "Fusion is easy," Edward Teller told them. "Control is hard."

"Take all the money you want," they said. "Take all the time you want. When can we have it running? Next year?"

Teller shook his head. "Maybe five years," he told them.

That was more than thirty-five years ago. Teller's first reaction was right: Fusion is easy, control is hard.

The fusion bomb is made by putting together a bagful of hydrogen isotopes and heating it to a temperature of millions of degrees at high pressure by exploding a fission bomb around it. For a microsecond this duplicates conditions inside the sun, with the end result that hydrogen nuclei fuse together, forming helium, and a blast of energy is released before the resulting shock wave blows the whole thing apart. It is often called a *thermonuclear* bomb, since the nuclear energy release is triggered by the heat of the fission bomb.

To produce a controlled, slow release of fusion energy, you obviously can't start with an atomic explosion. So somehow you have to find another way of heating the hydrogen to a million-degree temperature. And as soon as the scientists realized that, the next problem became clear: If you have something that hot, what do you hold it in? Any known material would vaporize at that temperature; in fact, they couldn't even *visualize* anything that wouldn't. And even if they did find a bottle to hold the hot hydrogen, the internal heat would be conducted to the walls of the bottle and be dissipated, lowering the temperature and shutting off the reaction.

The idea they came up with, which is one of the two solutions being sought today, is to confine the hot hydrogen *plasma** within a magnetic bottle, consisting not of a material substance but of magnetic lines of force so intense the charged hydrogen nuclei and electrons cannot escape. It's clever and sophisticated, and thirty-five years later it still doesn't work.

The other solution being chased is called *inertial confinement.* A fuel pellet of solid material is bombarded and heated either by a laser beam or a particle accelerator. As the outside of the pellet vaporizes it drives the rest of the fuel inward, accelerating it to high pressures and temperatures and, hopefully, thermonuclear ignition.

Will both or either of these techniques ever work? We don't know. Six years after our first fusion explosion, when the United States was planning its program for the Second Atoms for Peace Congress to be held in Vienna in 1958, Teller suggested telling the Russians all we knew. The AEC was astounded; Teller, after all, had been the AEC's leading advocate in the Oppenheimer trial, charging him with being a security risk. But Teller pointed out that the Russians had exploded their own hydrogen bomb less than a year after the Americans, and in some ways theirs was more advanced. Most important, he said, the American-controlled fusion program was stuck; they had made little progress in the past six years, and quite frankly didn't know where to head next. The problems of controlled fusion, in particular the magnetic bottle and the heating procedure, were totally different from anything applicable to building a better fusion bomb. It was a peaceful endeavor, and perhaps if we told all

*At this temperature the gas becomes ionized and consists of free nuclei and electrons; this is called a plasma.

we knew the Russians would reciprocate and between us we might make some progress.

The AEC reluctantly agreed, and since then the control of fusion has been an international effort. The magnetic-bottle design most actively pursued today is the Tokamak, a Russian idea, while the inertial-confinement concept is American in origin. There are three advanced Tokomaks currently operating: the Joint European Torus, the DIII-D at General Atomics in San Diego, and the PBX-U at Princeton University. The inertial-confinement machines are at the Livermore National Laboratory in California, the University of Rochester, and a new design at Sandia National Laboratory in New Mexico. In 1989 the news journal of the American Physical Society, *Physics Today,* reported that "extraordinary progress" had been made in the last two years, but the most we can hope for is that a laboratory small-scale demonstration and experimentation facility might be ready by the turn of the century.

It's been more than thirty-five years since the world's first thermonuclear explosion, since the AEC demanded a controlled fusion device "next year" and Dr. Teller told them it would take at least five years. It's been more than thirty-five years, and success is still at least another thirty-five years away.

At least. In 1975, at a series of lectures given in Israel, Dr. Teller was once again optimistic: "My optimism extends to the belief that in the next few years we will obtain 'energy breakeven' in laser fusion [inertial confinement]." "Energy breakeven" is the first threshold in obtaining fusion power; it means that a fusion reaction has been maintained long enough and at high enough power to put out as much energy as was put in to heat the hydrogen nuclei. After that, the *real* problems start: upgrading the experimental device to a commercially viable piece of operating equipment. Energy breakeven will be, as Winston Churchill said of the victory at El Alamein in 1942, "not . . . the beginning of the end, but perhaps the end of the beginning."

But we're not progressing toward breakeven quickly enough, because we're not trying hard enough. In 1986 the Electric Power Research Institute (the research organization of the commercial electric industry) cut out its $4 million annual funding for fusion. In the spring of 1989 the National Academy of Sciences pointed out that the United States government is currently assigning *half* as much funding to magnetic fusion as it did in 1977. The academy wants a 20 percent raise immediately, to build a Controlled Ignition Tokamak, and points out that electricity generation by Tokamak might be reached by 2025, but because of the industry's decision to divorce itself from fusion work, no company will be prepared to build a working Tokomak.

In 1975 Dr. Teller was hoping for breakeven "in the next few years." Fifteen years later, we're still hoping, still waiting.

And still dreaming. The latest dream was announced on March 23, 1989, when Stanley Pons of the University of Utah and Martin Fleischmann of England's Southampton University announced the discovery of "cold" fusion. The name comes from the claim that fusion had been accomplished without resorting to temperatures of millions of degrees. If true, all the problems associated with a controlled-fusion reaction would be licked, for they all stem from the difficulty of maintaining and enclosing the high temperature normally required.

Basically what you're trying to do in the fusion process is squeeze two hydrogen nuclei together, to bring them within range of the nuclear force. This is the strongest force in nature, but also the one with the shortest range; it dies out quickly at distances greater than 10^{-12} centimeters, about ten thousand times smaller than the size of an atom. The reason it's hard to do is that each hydrogen nucleus is positively charged, and identical charges repel each other with an energy that gets greater the closer the objects come to each other. Just before the two nuclei touch each other, at a distance just beyond the range of the nuclear forces, this *electromagnetic* repulsive force gets so large that it can be overcome only if the nuclei are hurtling toward each other terribly fast: at a velocity equivalent to a temperature of millions of degrees (since rising temperatures result in faster motion of the atoms and nuclei in the gas or plasma).

Once the nuclei reach this speed they can sail over the electromagnetic barrier into the range of the nuclear force, and when this happens they fuse together into a new nucleus since the attractive nuclear force is millions of times greater than the repulsive electromagnetic force.

But how can they reach the required velocity without the high temperature? There is in fact a reason behind the claim, which has to do with a quantum-mechanical concept called the "tunneling effect." To visualize this in everyday terms, imagine a wall ten feet high, and a baseball. If the ball is thrown at the wall, and is thrown more than ten feet high, it will pass over it; if it is thrown less than ten feet high, it will bounce back.

Simple enough. But in the quantum world of atoms and nuclei, things are not so simple. The analogue of the wall is the electromagnetic barrier created by the positive charges of the nuclei, and its height is measured in terms of their total repulsive energy. To carry out the analogy as the two particles are thrown together, if they have enough thermal energy to fly higher than the barrier they will fuse; if they have less than that energy, they will bounce apart. But in this queer quantum world there are third and fourth possibilities. Occasionally two particles with more than enough energy to fuse will still bounce apart, and occasionally two with less than enough energy will fuse.

In the everyday analogy, the former case would be like a baseball thrown twelve feet high and nevertheless "bouncing" back from the wall; the latter case would be that of a baseball thrown five feet high, hitting the wall, and passing right through it. This last effect is known as "tunneling" through the barrier, and it is a very unlikely but very real effect. It is responsible for the type of radioactivity known as alpha decay, in which alpha particles within a large nucleus continually bang up against an internal barrier and very occasionally tunnel through to the outside world; when they do, the atom is said to "decay."

In a normal hydrogen molecule there are two hydrogen atoms, each consisting of a nucleus and an electron. The two electrons bind the two nuclei together, at a distance much greater than the nuclear range of forces. It is possible that in their motions the two nuclei might tunnel through the electromagnetic barrier and fuse, but this is so unlikely a process that it is unobservable. The rate of tunneling, however, is a strong function of the "thickness" of the "wall": if the wall is made thinner, the rate of tunneling increases dramatically. In the hydrogen molecule, if the electrons are replaced by heavier particles called *muons,* the resulting muonic molecule is more tightly bound, the hydrogen nuclei are closer together, and the probability of tunneling through the barrier is greater. Thus "cold" fusion of muonic atoms actually has been achieved, but the rate of tunneling is still so slow that no significant energy can be produced; it's an interesting scientific phenomenon, but of no practical use whatever.

However, it leads to the possibility that if the two nuclei could be brought even closer together by some other means, the rate of tunneling might be increased. What Pons and Fleischmann did was to use palladium metal in an attempt to achieve this. It is well known that hydrogen diffuses into palladium, and can be concentrated there. If the concentration becomes high enough and the hydrogen atoms are squeezed closely enough together, might not fusion result?

Well, yes, it might, but again not frequently enough to be of any practical consequence. The rate of fusion that might be induced is many millions of times less than they have claimed to see. Their reports of heat coming out of the palladium apparatus have been both attacked and supported by other laboratories, but as has been pointed out, heat doesn't mean fusion. If you strike a match, you get heat out of it. Their claim for fusion is that they have got so much heat that it couldn't be anything else; most other scientists now feel that their measurements were mistaken, that there may be heat coming out but not as much as they claimed. In any event, if the heat had really been due to fusion, there would also have been so many neutrons emitted that they'd both be dead.

Their experiment was fun to read about and talk about, but it was such stuff as dreams are made on: dreams, not fusion reactors.

19

Bathtub Diplomacy

What is needed is a combination of humanity, common sense, and quiet confidence.
—Winston Churchill

Thermonuclear energy is not as beautiful a concept as we once thought. While it doesn't produce the intensely radioactive waste products that a fission reactor does, a fusion reactor does inescapably produce high-energy neutrons. These neutrons can be shielded and kept from escaping the reactor, but the shield then becomes radioactive. This is not a serious health hazard because the shielding material can be chosen to produce a low amount of radioactivity with a short half-life, so that disposal is not a problem, but it is something that must be considered and is an economic problem: the shielding material not only becomes radioactive but may become brittle as a result of the neutron irradiation, and must be continually replaced. And while normal hydrogen could be used as fuel for the reactor, the fusion reaction between the heavy isotopes deuterium (hydrogen-2) and tritium (hydrogen-3) has greatly enhanced capabilities. Given the difficulty in achieving a controlled reaction under any circumstances, it is clear that we will have to use the deuterium-tritium mixture as fuel instead of normal hydrogen. Deuterium is much scarcer than normal hydrogen, but is still plentiful enough for our needs, and tritium is produced as a by-product in the fusion process itself. The real problem is that tritium is radioactive. Although its radiation is very weak, it must be guarded against, particularly since escaping tritium can exchange isotopically with rain or ground water and would then be next to impossible to separate from it.

But these problems are not overwhelming; they can be dealt

with. If we can learn to control either fusion or solar energy and induce them to produce economically competitive electricity, they would clearly be the systems of choice and would last us well into the nearly infinite future. There is no overwhelming theoretical obstacle to either of these as energy sources, so far as we can see, and given Murphy's Law ("Anything that can happen, will happen") we have every reason to be optimistic that in the long run our energy problems will be solved.

The question is, do we *have* a long run? We are faced with the very real possibility of the collapse of our civilization or even of total extinction in the next century unless we can solve the problems posed by greenhouse warming, ozone depletion, and nuclear war. And each of these problems has the potential to flower into disaster more suddenly than we like to think. War, as we learned in the Serbian summer of 1914 and again in the Gulf of Tonkin some fifty-odd years later, can break out suddenly and escalate alarmingly before reasonable men have time to understand what is happening. And in the summer of 1989 a group of researchers at the universities of Copenhagen, Iceland, and Colorado presented evidence that the greenhouse effect may do the same thing. They found in Greenland ice cores evidence that the last ice age did not end with a systematic and gradual warming, but with a bang. In just twenty years the temperature and weather patterns changed from glacial to moderate, indicating that our atmosphere may exist in stable states and can flip from one to another without much warning. We might be in for just such a terrible flip instead of "merely" a gradual warming trend over the next hundred years, if we don't take action very quickly.

The ozone problem seems to be coming under control, if only we can convince China and the developing nations to forgo CFC production. Although they are at risk from increased ultraviolet radiation as much as anyone else in the world, it is unfair to ask undernourished and underenergized people to stunt their economic development unilaterally for a benefit that will be shared equally by all of us. But it is within our power to offer economic inducements to offset the costs of their voluntarily not producing CFCs. No such economic inducement comes free, of course; we will pay for it in one way or another. Ultimately this means a lower standard of living for those of us in the "advanced" countries. But if it is borne equitably by all the advanced countries the decline per person should be immeasurably small, particularly when compared to the incredible burden we all voluntarily (or at least passively) bear which goes by the name of "defense."

Defense spending at some level is necessary; there is no point in being either good or wealthy if you are not strong enough to defend your goodness and your wealth from those who would obliterate the one and take the other. But surely we have been overdoing it for the past few decades. Since the Second World War we have spent on the order of $10 trillion on defense, to make ourselves safe, and it hasn't worked and it cannot work.

The concept that emerged out of the mushroom cloud at Hiroshima was fear of and reliance on the awesome power of the atomic nucleus. It was possible to build an arsenal of atomic bombs, and aircraft and missiles to deliver them, that could "reduce all of Russia to a smoking, radiating ruin" within two hours. And so this arsenal was built and we huddled behind it and told ourselves we were safe, for who could attack a nation with such an overwhelming capability?

The answer is, anyone who's insane. And in a century that has seen the insanities of Hitler and Stalin, who can doubt that insane people can reach positions of power in political organizations? We are each of us in imminent danger of someone, somewhere, pushing the button: if anyone does, nothing in our power can stop the missile or the smuggled bomb. The only thing our nuclear arsenal protects us against is an all-out attack by reasonable people. But before we developed and built that arsenal, we were just as safe. We were surrounded by impassable oceans on two sides and by friendly and subservient neighbors on the two others: no nation on earth could have attacked our homeland. What have we gained defensively for our $10 trillion? Not much.

We have gained just as little in offensive capability. Our own power of atomic destruction has turned out to be just too powerful to be used. Here we sit with all kinds of strategic nuclear weapons and all kinds of missiles, the strongest people that ever existed on the face of the globe: we can actually destroy anyone, anywhere in the world, anytime we want to.

And yet we can't. It's like being in a room with a pesky fly and a flyswatter. You can kill the fly with the swatter, but you can't make him behave; you can't make him do what you want him to do. All you can do is kill him—and we can't even do that, because we live in a kind of Society for the Prevention of Cruelty to Flies: world opinion.

So what can we do with our nuclear weapons? We couldn't stop the Vietminh from marching south to Saigon, and we couldn't stop Russia from taking Czechoslovakia. We couldn't stop Syria from attacking Israel, and we can't stop terrorists from blowing up our airplanes and kidnapping hostages. We couldn't even stop Noriega, the Sandinistas, or the Cubans.

We could swat them, sure; we could kill them all. But we can't make them behave the way we want them to. So what good is our nuclear-deterrent capability? What sort of antisocial behavior can it deter? One thing only: it can stop another country from launching an all-out nuclear attack on us. It's a weapon whose only use is to prevent its ever being used. Because everyone knows we're not going to launch our nukes every time Iran mines a harbor or the PLO kills a busload of children or the Soviets march into Poland or the Chinese into Tibet.

Which is not to say we were wrong in developing it. We were then in the midst of a terrible world war, and the secret of nuclear fission had been discovered by our enemies: nuclear fission was first accomplished in a laboratory in Fascist Italy, and first recognized by scientists in a laboratory in Nazi Germany. There was every reason to think that the Germans would be ahead of us in the effort to develop an atomic bomb, and whichever nation had the bomb would clearly win the war. There is even a good argument to be made for our use of the bomb when Germany had already surrendered and it was clear that the Japanese were not an atomic threat and had in fact already lost the war. Although the outcome of the Second World War was not in doubt by the summer of 1945, everyone agreed that there was a lot of bitter fighting still ahead, with casualty estimates running as high as one million Allied lives and twice as many Japanese in the battle for the Japanese homeland. There is a good argument to be made that the bomb was a useful and even necessary weapon in that war, ultimately saving many lives, perhaps millions.

But what has it done for us lately? It has led us into the valley of the shadow of death, and our proud and stupid boast has been that we fear no evil because we are the toughest sons of bitches in the valley, armed with a weapon that we dare not use.

Today more than ever we dare not use that weapon, for reasons stronger than fear of world opinion and more imperative than questions of morality: we dare not use it because it would be suicide.

In the early 1950s Dwight Eisenhower seriously considered attacking Russia with our full arsenal of nuclear weapons. In the early 1960s Jack Kennedy brought us to the brink of nuclear warfare with the Cuban crisis. In the late '60s Barry Goldwater, running for President, wanted us to bomb the North Vietnamese "back into the Stone Age." At none of those times did anyone even suspect that use of our nuclear arsenal on an enemy would have propelled the entire Northern Hemisphere and perhaps the entire world into at least a nuclear autumn if not a nuclear winter. Without knowing it, we could have brought down upon our own heads a calamity worse than any civilization has ever endured. Never have the biblical words rung more true: "Father, forgive them; for they know not what they do."

Over the past several decades we have learned that our terrible nuclear strength is of no avail against guerrillas or terrorism, and now that we understand the consequences of and the possibility of nuclear winter and ultraviolet summer we have come to learn that our nuclear weapons aren't even useful for a major conflict unless we are prepared to face the very real possibility of mass suicide.

It's difficult to make a screeching stop and reverse direction. It's hard enough on a bicycle; it gets progressively harder on a motorcycle, in a sports car, a truck, an intercontinental ballistic missile, and finally in an entire nation. It's hard, but when compelled by the facts it is sometimes necessary.

The possibility of nuclear winter means that an obliterating nuclear attack—the overwhelming preoccupation of both our defensive and offensive strategies for the past forty years—would be suicide, so who would launch it? No one but a maniac, and the best preventive for such an occurrence is not a reasonable deterrent—for maniacs are not convinced by reason or by deterrents, as we learned in 1939—but the removal of all such possibility, doing away with nuclear weapons.

Our entire defense posture must be revamped. This is not to say that all nuclear weapons should be dumped immediately, because a deterrent must still be maintained against any small country that puts together a bomb and threatens to nuke an American city. But since the idea of our launching fifteen thousand nuclear warheads is no longer feasible, there's not much point in our keeping those fifteen thousand warheads ready to fly at the touch of a button.

Getting rid of them does not represent a major cut in immediate defense expenses, but it's certainly going to help. Consider, for example, tritium. As it is in the controlled fusion process, tritium is one of the components of the bomb. And tritium is radioactive, with a half-life of twelve years, which means that the tritium base of our nuclear bombs is continually decaying and must be replaced on a time scale of decades. This is a major factor in the $4 to $8 *billion* the Department of Energy wants to spend on new defense reactors, a figure that will surely grow to $10 or $20 billion as all defense contracts do. That money could be used on energy research, to find the proper fuels and procedures that will solve the greenhouse problem. We are currently spending $300 million on fusion research and less than that on solar, so an increase in the pot of several billions would really be significant.

It actually comes to quite a bit more than that. A censored version of a report that the DOE sent to the White House in December of 1988 reveals, half-hidden in a morass of words and figures, that over the next twenty years it will cost us $244 *billion* to keep our nuclear-weapons factories. The major portion of this money is for waste disposal and environmental protection, two issues that the defense establishment has been postponing and which now have reached levels impossible to ignore. Hazardous wastes have been stored above ground or nearly so at the Hanford and Savannah River military complexes, in leaky containers that would never have been countenanced for a moment in civilian reactor facilities.

Those groups arguing against nuclear power for commercial purposes in the United States have not realized that, like it or not, we have been a nuclear nation since the bomb first went off. It is just that all the radwaste and all the problems that go with giant reactors have been for the benefit of the military instead of the civil population. And because of that the problems have been kept free of regulatory interference on the grounds of military secrecy, to the point where they now represent significant health hazards that will cost more than $200 billion to clean up.

That money is going to have to be spent, but if we can agree that we don't need to keep producing more and more nuclear weapons we can not only prevent such potentially disastrous radioactive hazards from recurring, we can spend nearly half of that $244 billion on other programs, including the work that will help us to understand our planet and what is happening to it before it all blows up in our faces.

If we can manage to do that, perhaps we can also manage to redirect our thinking. The American Institute of Physics in 1986 compared the salaries of bachelor's-level physicists working for defense contractors with those working as teachers: the defense industry was paying a median monthly salary of $2,315, the education system was paying $1,350. If we pay our weapons designers nearly twice as much as we pay our teachers, we deserve what we get: better weapons and dumber children. Is this the reason that in 1987 for the first time in history the list of the top three industrial recipients *for United States* patents did not contain the name of a single American company? Instead of General Electric and RCA and IBM, the names were Canon, Hitachi, and Toshiba.

The only argument against a drastic reduction of our nuclear deterrent capability is that it is needed to guard against a massive nuclear attack, and that means from Russia or China. The other nuclear nations have vastly smaller numbers of missiles, and no air force or ground armies that could threaten our security. Not only does nuclear winter remove the threat of a Chinese or Soviet attack—unless we are seriously worried that either of these nations might commit suicide in order to get rid of us—but the greenhouse effect and the ozone-depletion danger make it imperative that we begin serious joint research and cooperative action plans with these nations as well as with the smaller nations of the globe.

There are many people who argue against any form of cooperation with an "enemy." The basic line of reasoning is that what is good for them must be bad for us, and conversely that for us to prosper they must suffer. But as Ira Gershwin tried to tell us, it ain't necessarily so. I would like to offer an alternative scenario, which might be called bathtub diplomacy, and which can be illustrated by a simple anecdote.

A few winters ago I was in England attending a scientific conference, and took advantage of an empty weekend to travel to Chichester, primarily to attend the Chichester Festival's production of John Arden's *Armstrong's Last Goodnight.* I stayed at a seventeenth-century inn, which was remarkably comfortable except for the absence of much heat. The weather was abominable, cold with wind and rain, and the small gas heater in the room was nearly useless. There was, however, in the inn's one bathroom an enormous old tub, so large that when I reclined with my head against one end of the tub I could not reach the other end with my toes. Unfortunately the hot-water supply was not equal to the dimensions of the tub, and it was impossible to fill it with enough hot water to cover my body. It came tantalizingly close, but no matter how I scrunched down, the top of my chest and belly remained out in the cold air.

On Sunday afternoon, after a chilling walk through the town, I returned to the inn cold and wet, soaked through by the drizzling rain, and reached the bathroom door at the same moment as a young lady who had approached from the other end of the hall. We stood hesitantly before the door, each knowing full well there was scarcely enough hot water for one bath, with hours to wait before the boiler cooked up enough for a second. I thought of suggesting that she go first but refrain from using all the hot water, but that would give her a miserable bath. I could have gone first myself under the same restriction, but with the same result; in addition I didn't trust myself not to use all the hot water, which meant of course that I couldn't trust her.

The same thoughts were clearly going through her own mind. It could have been a dreadful situation, until one of us suggested that though we were competitors for the underabundant hot water we could cooperate with benefits to both. And so we shared the hot bath, and found that not only was there no less hot water for each, there was actually more. For the two bodies in the tub raised the water level sufficiently for it to cover both adequately.

The story is without doubt apocryphal, but it illustrates the benefits of cooperation, even between competitors.

20

The Power and the Glory

Do you think, because I am poor, obscure, plain, and little, I am soulless and heartless? You think wrong!

—Jane Eyre, by Charlotte Brontë

We are a democracy, but old habits die hard, and we the people have a tendency to sit back placidly and wait for the great ones of this world to solve our problems. We tend each of us to inhabit our own little universes, and to think of the big one beyond our individual skins and our immediate families to be too vast, too powerful for us to influence. We muddle through our own private lives as best we can, glancing only occasionally and fearfully at the problems of the larger world, having faith that someone stronger, more clever, more powerful than we will surely solve the problems that vex us. We have faith in our own particular gods, and turn to them in times of trouble and danger. In the winter of 1988–89, citrus growers in South Texas who were still recovering from a drastic freeze in 1983 were warned of another spell of bad weather on its way toward them from Canada. They took suitable precautions: "As people have been leaving my store today I've been telling them to go to church Sunday and pray for us," Mandy Eggers told reporters from her family's Rio Grande Valley farm.

More sophisticated people share that same naïveté; they simply pray to a different god. Professor S. W. Leslie of Johns Hopkins University bemoaned this tradition in the journal *Science,* arguing that "Over the years Americans have shared an intense faith in technology as the efficient solution to our messiest social problems." And there are a lot of high-tech solutions being talked about out there, and someday they may be implemented. Today they're just things that are fun to talk about, of not much more use than prayer.

Roger Revelle, who was one of the first scientists to recognize the greenhouse problem, has suggested using the photosynthetic potential of phytoplankton. These are small marine plants that drift on or near the surface of the oceans, and like all plants use sunlight to combine carbon dioxide and water into carbohydrates. In the normal course of events they are eaten by fish, but if they should die and sink before they were eaten, the carbon they removed from the atmosphere would remain on the sea floor for hundreds of years. We could conceivably accelerate their photosynthetic activity by fertilizing the oceans to produce large plankton blooms, so large that when they die they would sink before being eaten by their natural predators, thus shifting carbon dioxide from the atmosphere to the ocean bottom.

Michael McElroy, an atmospheric scientist from Harvard University, has suggested a sexier system: we could launch into space orbit large reflectors, which would beam winter sunlight at Arctic and Antarctic oceans, which are normally in the dark during winter. This extra sunlight would stimulate photosynthesis to give the same effect as Revelle's proposal. (One might also use such a system of orbital mirrors to deflect sunlight away from equatorial regions and out into space, thus cooling the earth directly.) Wally Broecker, a geochemist at Lamont-Doherty, wants to use a fleet of seven hundred jumbo jets to spread thirty-five million tons of sunlight-reflecting sulfur dioxide around the stratosphere. Thomas Stix, of Princeton, says we could remove the CFCs from the atmosphere by taking advantage of their strong infrared absorptivity: we could actually zap them with IFR lasers and blast them out of the atmosphere.

All these solutions are possible in the future, but all of them have problems for the present. We don't yet know enough about the food cycle in the oceans to be able to predict quantitatively the effect of fertilizing enough of the ocean to make the phytoplankton solution tenable. There are similar problems in throwing all that sulfur dioxide into the stratosphere, and obvious engineering problems in orbiting a fleet of mirrors or building and operating enough lasers to sweep the air clean. Someday we may be able to do such things, but there are dangers in rushing in with large-scale technological solutions before we're sure we know what we're doing.

Take the blackflies, for example. In Maine the tourist fishing industry is hurt every summer by these nasty biting pests, which make a lot of people swear never to return. After World War Two the local chambers of commerce had a wonderful idea: DDT had just been released for civilian use, and was the marvelous technological answer to all humanity's insect problems. So they sprayed the rivers and practically the whole state, and the blackflies were nearly exterminated. Everyone had a wonderful summer.

But within a few years the blackflies were back and were more numerous and more terrible than ever, because they had been only *nearly* exterminated, and their natural predators had fared even worse. We didn't really know what DDT would kill, and as it turned out it killed the creatures who fed on the blackfly eggs more efficiently than it killed the blackflies. The next few generations of blackflies were fewer in number, but they spread like—well, like blackflies. Without the natural predators to keep them under control, they were fruitful and multiplied and covered the earth, ruining the tourist business for several years until the natural balance was slowly restored.

So we have to be careful. Before we start messing with the oceans and the atmosphere we have to learn more about them. We have to start using the money we can save from military nuclear-research programs to fund basic studies of the oceans and atmosphere. This is necessary before we begin even more prosaic wide-scale solutions, like shifting the courses of rivers to deflect warm waters from flowing into the Arctic or cold waters from flowing out, or even the seemingly simple expedient of painting rooftops white all over the world, to reflect away more sunlight. The questions remain: How much is enough? How much is too much? Today, we just don't know; we don't know enough about any geocycle on earth to be able to roll up our sleeves and get to work.

The things we *can* do at the present time are provide more money for basic research, conserve our use of fossil fuels by making our engines and buildings more efficient, and begin a cautious but definite move toward nuclear power. None of us as individuals can do very much in these regards, but we can do a lot if we join the political system. Because the power and the glory of a democracy are also its greatest weakness: it responds and adjusts to the will of the people, but does nothing if the people don't care.

In the long run our gods and our scientists may come round to save us, but to make sure we're still around when they find the solutions we have to rely on our politicians. They are the ultimate enemy and also the ultimate savior, for as Pogo has said, "We have met the enemy, and they are us." If we join forces and push our governments in the proper directions, we have within our means the power to slow down and halt the depletion of the ozone, the increase of the greenhouse effect, the threat of nuclear winter. We have this power *today,* and it has to be used right now.

There are a myriad of political organizations already in existence; their impact depends on the support they get from the people. The Wilderness Society, Greenpeace, the Environmental Defense Fund, the National Resources Defense Council, and the Audubon Society are concerned with the environment, SANE with safe limits to nuclear weaponry, the Center for Defense Information with war and defense postures, Common Cause and the Worldwatch Institute with all of the above. A list of these organizations and their missions and addresses is found in the Appendix.

Wilt thou Go with Me and do battle?

The final paradox of these coming catastrophes is that no matter what action we take we can't solve the problems alone; and that fact becomes both our greatest peril and our ultimate hope.

One of the oldest clichés of science fiction is the story of the evil genius or the swarm of extraterrestrial aliens who threaten the destruction of the world; it is a plot which, if handled deftly, never fails to thrill. It is even, and with some reason, taken seriously by serious people. There have been arguments made against a scientific search for extraterrestrial intelligent life forms, on the basis of what our own "superior" societies did to more primitive peoples when our explorers discovered them and on the fear that superior alien beings might do the same to us. On the other hand, others have speculated that only through some such overwhelming encounter will we rise up out of our petty difficulties and confront the real problems that surround us.

The postulated encounters with extraterrestrials are generally of two kinds. In one we are touched by a benevolence amounting almost to the godly, in the other we are confronted with an evil so powerful that it is overcome only by all the nations of the earth uniting in opposition. In the latter scenario the happy ending comes about not only with the alien defeat, but with the resulting knowledge that if we all work together we can defeat any threat.

This is a theme that we have seen expressed in the real world as well as in fiction. For centuries Europe dominated the globe and at the same time continually fought terrible internecine wars. Today Europe's economic existence is threatened by the United States, Russia, and Japan, and as a result the European nations have banded together into the European Economic Community. It would be unthinkable today for Great Britain to go to war with France, or either of these to march against the German Federal Republic. The dream of centuries has become reality—not through brotherly love, but out of stark necessity.

This is the key to our future: not necessarily to love our neighbors, but to realize that we need them and for them to understand that they need us. For what will it matter that we outlaw all CFCs, if the Chinese begin to use them? Today the Western nations account for 60 percent of the world's fossil-fuel burning; if tomorrow our contribution drops to zero but the developing nations triple their energy demands, the global carbon dioxide will continue to climb inexorably toward catastrophe. If any one nation on earth begins an all-out nuclear assault, the long winter will descend on us all.

The script being thrust on us today reads almost as if it had been written by a heavenly sci-fi author. Earth is threatened by destruction on a world-wide scale, and only if we all work together can we defeat the forces of evil, which are no less evil for being impersonal: drought and flooding, ultraviolet radiation and smog, cancer and famine, all brought on by the three horsemen of the apocalypse—the greenhouse effect, ozone depletion, and nuclear winter—riding the singular steed of atmospheric alteration.

We realize finally that we live at the bottom of a protective ocean of air, and that this atmosphere is held in common by all of us. As all of us must suffer if any of us degrade it, so all of us must work together to rescue it from harm. We must not make the mistake of thinking that because the small, undeveloped nations of the world are by comparison to ourselves poor, obscure, and little, they are also soulless and heartless. The dangers confront and concern us all, and are real, immediate, and probably solvable. More than that, it is the threat of these dangers that will in the end save us, because for the first time we confront disasters that may befall all of us on this earth, that are easily understandable and have consequences immediately apparent to everyone, nomads and farmers and hunters and city dwellers alike, from east to west and from pole to pole. We all see the deserts expanding and the fresh water diminishing, the temperatures rising and the rains not coming, the air choking us, the forests and wildlife dying.

For the first time we have a threat that affects the entire world and that is frighteningly demonstrable today. For the first time we have the opportunity to gather together and work as a united people against a danger that is real, here and now, and affects us all. Up till now we in the United States have been ineffectual leaders of the world's teeming populations; they sit back, remote and uninvolved as we build more and more missiles, as we race to the stars. They do not admire our civilization but rather resent us as unfair usurpers of the world's wealth; they do not join in our endeavors, but see us as alien creatures, powerful and selfish, clever and self-indulgent, technically dominant and morally bankrupt. But now we have the opportunity to lead people who will follow because they are concerned about the fight, because it is their fight as well as ours.

All the problems we have talked about come together here. The promise of nuclear energy will enable us to cut down on our use of fossil fuels, thus diminishing the greenhouse effect—if we can solve the problems of nuclear safety and security against terrorism. The restoration of the ozone layer will shield us from the sun's ultraviolet radiant energy, and the reforestation of the world's jungles and wooded areas will once again provide the natural sinks for atmospheric carbon dioxide that are needed to balance the fuel burning, if the requisite international and monetary cooperation can be achieved.

Contrary to research on nuclear and missile systems or space science, the kind of research that is necessary to solve the atmospheric problems can and must be done in the developing countries as well as in our sophisticated research centers. Large expenditures and giant national laboratories are not enough; on the contrary, a large part of what is needed is field work in the areas of the world most important to global climate, and many of these geographical regions are in undeveloped parts of the world such as the Congo and

Amazon basins, the tropical rain forests, and the deserts and rice fields of the world.

We can do this, we can do it all; it's not too late. The catastrophes are coming, but they're not upon us yet. The game, as they say, isn't over until the fat lady sings, and she hasn't started singing yet.

But if you listen carefully you can hear her offstage, in the wings, beginning to warm up.

I imagine that the future is uncertain. And because it is uncertain, I am determined to do something about it.

—Edward Teller

When your brain and heart are engaged,
you can't go wrong.

—Yo-Yo Ma

Appendix

The *Environmental Defense Fund,* 257 Park Avenue South, New York, NY 10010, is currently raising funds to apply pressure on the Brazilian government to arrest and prosecute the people who are murdering those who have spoken out against destructive land-use policies there, to influence the World Bank to hold to its environmental promises, and to stop governmental subsidies and tax breaks which subsidize the destruction of the rain forest.

Common Cause, P.O. Box 220, Washington, DC 20077-1275, is a citizens' lobbying organization which focuses on the governmental process. Priorities for action are decided upon by membership surveys and polls. Recent issues have included energy conservation, conflict of issues and financial disclosures relevant to military funding and procurement, ethics in government, and the dangers to America posed by our military programs.

The *International Oceanographic Foundation,* Rosenstiel School of Marine and Atmospheric Sciences, University of Miami, Miami, FL 33149, is devoted to raising funds for oceanographic research and to promoting public interest in and knowledge of oceanic resources.

The *National Audubon Society,* P.O. Box 51000, Boulder, CO 80321-1000, is concerned with pollution, particularly of the wilderness; two of its primary emphases are the need for energy conservation and legal penalties for polluters.

Greenpeace, 1436 U Street NW, Washington, DC, 20009, is an international environmental group which organizes and lobbies on environmental and disarmament issues. It is one of the most aggressive organizations (its ship, the *Rainbow Warrior,* was sunk by the French secret service while it was protesting France's nuclear testing in the Pacific) and one of the most effective, although I think its antinuclear stance carries it too far against civilian nuclear energy.

The *Worldwatch Institute,* 1776 Massachusetts Avenue NW, Washington, DC 20036, recently has been concerned with the hidden costs associated with nuclear power, the spoilage of the oceans, the high cost and inefficiency of the military program, solar power, and the mismanagement of our toxic waste cleanup program.

SANE, 5808 Greene Street, Philadelphia, PA 19144, is dedicated to examining the nuclear alternatives available to us, with particular emphasis on ending the nuclear arms race.

The *National Resources Defense Council,* 40 West 20th Street, New York, NY 10011, addresses a wide range of environmental problems. Recently it has been involved in developing programs to protect against the greenhouse effect through energy conservation, in helping the American and Soviet governments to evolve a workable nuclear test-ban verification scheme, in protecting our forests and water systems, and in working with Congress on clean air legislation.

The *Wilderness Society,* 1400 Eye Street NW, Washington, DC 20005, is concerned with defending America's national forests and other wild lands. This includes protesting against the U. S. Forest Service as well as against industrial sources of pollution and deforestation.

The *Center for Defense Information,* 1500 Massachusetts Avenue NW, Washington, DC 20005, is an organization directed by former military officers, devoted to finding alternatives to war.

Finally, some of the most useful organizations are those formed by preexisting groups, such as *Physicians for Social Responsibility* and the *Union of Concerned Scientists.* If your profession/union has such a group, you might want to begin there.

Notes

(References are to the Bibliography. Citations without authors are news articles.)

Chapter 1: Fire

"No more politicians. . ." Paul West, as quoted in Littmann and Yeomans. All anecdotes about the comet are from this book.

"Dinosaur history. . ." *Science News,* 10/4/86, and Encyclopaedia Britannica (see under Mantell).

"All of Russia. . ." Powers.

"You are alive at this moment. . ." and "ten trillion dollars. . ." Lecture by C. Sagan at the New Jersey Chapter of SANE Freeze, broadcast on "Consider the Alternatives."

Chapter 2: Ice

"Helium-3. . ." To be precise, the ratio of helium-3 to the more abundant helium-4 isotope can be as high as 0.35 in meteorites but is 0.000001 or less in terrestrial rocks. We found a ratio of 0.3 in the Washington County iron.

"Harold Urey. . ." from Craig et al. and Emiliani.

Chapter 4: The March of Ice

"Milutin Milankovitch. . ." From "V. Milankovitch: The Memory of My Father," in Berger et al.

"The tilt of the earth's axis. . ." Levin; Smoluchowski.

"The amounts of carbon dioxide dissolved in the oceans. . ." *Science,* 1/29/89.

Chapter 5: The End of the World: Part 1

"Man did not exist. . ." Flint.

"Frankenstein. . ." Gribbin; Schneider; Stommel and Stommel.

"Astrology poll. . ." *Science,* 1989.

Chapter 6: UV-B

"Geoffrey de Havilland. . ." Hallion.
"McDonald and the UFOs. . ." Dotto and Schiff, p. 43.
"SST advisory panel meeting. . ." Discussion from Dotto and Schiff, pp. 46–58.
"Discovery of CFCs. . ." Brodeur; Snell.

Chapter 8: The End of the World: Part 2

"Carbon dioxide in the air (historical record). . ." Bernard.
"Figure 8-1. . ." Data from Clarke and *Science,* 2/17/89, p. 891.
"One attempt was made. . ." See Bernard, and Broecker.

Chapter 9: The Ghost of Climate Future

"Box models. . ." Revkin, 1988.

Chapter 10: The Long, Hot Summer

"Cutting down the forests. . ." Adams et al., and *Science,* 9/30/77.
"Stratospheric Sudden Warming. . ." McGuirk.
"How much the waters will rise. . ." Bolin.

Chapter 11: A Backward Glance into the Future

Opening quote. . . *Science,* 1/13/89.
V. Ramanathan and R. Revelle quotes. . . *Time,* 10/19/87.
J. Hansen quote. . . Revkin, 1989.
"In May 1989. . ." *Science,* 6/2/89.
"Variation in warming from place to place. . ." *Science,* 2/5/88.
"Hot weather has always been with us. . ." Quotes from Miami *Herald,* 10/18/88.
"A few tenths of a percentage point. . ." and following quotes from Miami *Herald,* 10/18/88.
"Be fruitful and multiply. . ." Genesis 1:28. The Jerusalem Bible, Doubleday, 1966.
"In 1987 Donald Blake. . ." Bishop.
"Even worse is the estimate. . ." *Science,* 11/23/84.
"The only thing we are sure of. . ." Clarke.
"Early estimates of methane production. . ." Bishop.
"Last year the Institute for Space Research. . ." *Worldwatch,* Jan.–Feb., 1989.
"Although the methane produced this way. . ." Craig et al., 1988.
"And the government did not prosper. . ." and "If we were simply to let these uneconomic forests stand. . ." *Science News,* 6/4/88, quoting World Resources Institute economist Robert Repetto.
"One further methane worry. . ." Bishop.
"The models have a wide range. . ." Clarke.

Chapter 12: The Hole at the Bottom of the World

"The price of liberty. . ." Often attributed to Thomas Jefferson. But see Bartlett's *Familiar Quotations.*

"Early in 1976. . ." The group decided to check on the atmospheric stability of chlorine nitrate ($ClNO_3$), a compound formed by reaction between chlorine oxide (ClO) and nitrogen dioxide (NO_2). Chlorine oxide forms part of the chlorine chain that destroys ozone in the stratosphere (Chapter 7), and nitrogen dioxide is one of the nitrogen oxides that also destroy ozone as part of the natural background effect. If these two compounds were to get together to form chlorine nitrate, they would obviously be diminished, one molecule of each being lost for each molecule of $ClNO_3$ formed, and the result would be less of an ozone depletion than the Rowland theory predicted. A couple of decades previously (before the ozone controversy had been thought of) a group of German scientists published a paper showing that the $ClNO_3$ molecule would be broken up by sunlight, changing back to ClO and NO_2, and thus the atmospheric concentration of these two would not be affected. But when Rowland's group reinvestigated the problem they found some discrepancies in the German work and finally concluded that $ClNO_3$ would be reasonably stable after all, sufficiently so to reduce the earlier estimates of ozone depletion by about 20 or even 30 percent. Dotto and Schiff, pp. 243–45.

"du Pont asked the public. . ." *Science News,* 9/11/76.

"In September a National Academy of Sciences panel. . ." *Science,* 10/8/76.

"Russel Peterson. . ." Dotto and Schiff, p. 282.

"The National Research Council recommended. . ." and "This was about half the effect. . ." *Science,* 10/8/76.

"By the spring of 1977. . ." Dotto and Schiff, pp. 287–88.

"They couldn't do much else. . ." Dotto and Schiff, p. 297.

"A 16 percent reduction. . ." *Science,* 1/25/80.

"The response was that. . ." *Science,* 12/7/79.

"They jumped up from their chairs. . ." All right, I dramatized this part a bit.

"The British Antarctic Survey. . ." Gribbin, 1988, pp. 107–12.

"Susan Solomon at NCAR. . ." *Time,* 10/19/87.

"Measurements from Nimbus 7. . ." *Newsweek,* 6/23/86.

"And in Tasmania. . ." *New Scientist,* 5/27/89.

"In December the United States. . ." Mintzer.

"Twenty-eight nations met in Geneva. . ." *Science,* 5/29/87, and Clarke.

"The cost of those lost lives. . ." *Science,* 5/29/87.

"A variety of nonpropellant functions. . ." Snell.

"CFC-22. . ." *Nature,* 2/9/89.

"EPA seems happy with the protocol. . ." Miami *Herald,* 10/19/88.

"Margaret Tolbert. . ." *Science News,* 3/19/88.

"A $25 million plant in Texas. . ." *New Scientist,* 10/29/88.

"March 1, 1989. . ." Miami *Herald,* 3/3/89 and 3/4/89.

"And what has the delay cost us?. . ." Woods.

Chapter 13: The Black Cloud

"A bunch of the boys were sitting. . ." Turco et al. and *Time,* 12/24/84.
"But rain occurs only. . ." Much of this and the following are from Turco et al.
"Our current Strategic Integrated Operating Plan. . ." Powers.
"Subjected to prolonged darkness. . ." This and following quote from Turco et al.
"Joseph Knox. . ." Turco et al.
"Enough NO_x could be produced. . ." *Science,* 12/23/83.

Chapter 14: The Sky Is Falling

"The idea of nuclear winter was invented. . ." *National Review,* 12/19/86.
"In a previous article. . ." *National Review,* 11/7/86, quoted in *National Review,* 12/19/86.
"Despite the fact. . ." *National Review,* 12/19/86.
"Physical and biological scientists. . ." Schneider.
"An anonymous MIT professor. . ." The Soviet physicist, Vladimir Aleksandrov, has since disappeared under mysterious circumstances, and so can't be queried about the extent of and reasons for his disagreement. See Sparks.
"In the fall 1986 issue. . ." *New Scientist,* 12/11/86.
"In subsequent interviews Dyson. . ." *New Scientist,* 12/11/86.
"Nuclear winter isn't science. . ." This and following quote from Sparks.
"Sagan conducted an end run. . ." *National Review* editorial, 12/19/86.
"Ultraviolet summer. . ." Levi and Rothman.
"The consensus of the meeting. . ." *Science News,* 9/22/84, p. 182.
"Nuclear Winter is no fantasy. . ." Chown, *Nuclear War.*
"The Pentagon immediately repudiated. . ." Chown, "Smoking. . ."
"A preliminary three-dimensional model. . ." The quote is from a Thompson editorial, quoted in Schneider. Other material in this and the following paragraph are from Schneider.
"As Thompson and Schneider have said. . ." *New Scientist,* 12/11/86, and Schneider.
"In the summer of 1989 Jenny Nelson. . ." Nelson.

Chapter 15: The Legion of the Lost

"A measured concentration about three times higher. . ." Krauskopf.
"V. G. Gorshov claims. . ." Quoted in Kondratyev, p. 12.
"K. Ya. Kondratyev. . ." Kondratyev, p. 12.
"When ice melts. . ." Kondratyev, p. 22.
"The CO_2 climate effect. . ." Kondratyev, p. 62.
"The ocean will also get warmer. . ." McLean.
"The military research and development budget. . ." *Science,* 5/8/81.
"In the early 1950s. . ." *Science,* 7/8/83.
"We don't want to scare the country. . ." *New Statesman,* 1/25/85, p. 15.
"All of Russia would be nothing. . ." Powers.

Chapter 16: The New Conservatism

"Without improved efficiency. . ." *Science News,* May 7, 1988.
"This small cutback. . ." This and other facts following from Lovins and Lovins.
"In 1989 the city of Los Angeles. . ." *Science,* 2/17/89, p. 896.
"Professor John Blackburn. . ." All the foregoing arguments from Raloff.
"In March of 1977. . ." *Science,* 4/1/77.
"Two million square kilometers of forest. . ." *Science,* 12/16/88 (Letter by G. M. Woodwell).
"This loss of forest lands. . ." *Science,* 11/14/86.
"Applied Energy Services. . ." *Science,* 10/7/88, and *Science News,* 12/24/88.
"To absorb five billion tons of carbon. . ." *Science,* 10/7/88.
"In 1988 the American Forestry Association. . ." *Science,* 10/21/88.
"Global Re-Leaf Project. . ." *Science,* 10/28/88.
"We could do even more. . ." *Science News,* 5/7/88.
"Students for the Children's Rainforest. . ." Miami *Herald,* 3/5/89.
"Several legislative acts. . ." *Nature,* 3/2/89.

Chapter 17: The Second Coming

"Killer smogs. . ." Fergusson.
"In order to make a nuclear explosion. . ." Description from McGraw-Hill Encyclopedia of Science (under Atom Bomb).
"Niels Bohr. . ." and "the amount of radiation people get. . ." from Teller.
"Early in the morning of April 26. . ." Wilson; Snell; and *Science,* 12/16/88.
"Why did Chernobyl happen?. . ." Details from *New Scientist,* 9/4/86, and R. Wilson, "A visit to Chernobyl," in *Science,* 6/26/87.
"The yearly Doublespeak Award. . ." *Science,* 12/7/79.
"In 1977 an AEC report. . ." *Science,* 4/1/77.
"In 1989 the Institute for Energy. . ." *New Scientist,* 5/13/89.
"A graphite-moderated, water-cooled reactor. . ." *Science,* 5/1/87.
"But despite all this official secrecy. . ." Hoyle and Hoyle.
"The only incident of its kind. . ." There have been deaths from experimental reactors, but even these are measured in single digits. See *Science,* Letters, 7/13 and 11/9, 1979.
"March 18, 1937. . ." Miami *Herald,* 3/18/89.
"The Macchu Dam. . ." Hoyle and Hoyle, pp. 33–36.
"In the past 30 years. . ." *Science,* 4/1/77.
"The Non-Proliferation Treaty of 1968. . ." Patterson.
"In 1967 our AEC. . ." Patterson, p. 74.
"In that same year. . ." Patterson, p. 68.
"An old Indian fable. . ." *Science,* 8/13/82.
"We have enough uranium. . ." and "Are breeders the answer?. . ." Weinberg.
"We are a Christian nation. . ." Miami *Herald,* 3/16/89.
"The Super-Phénix has had more problems than anticipated. . ." *Physics Today,* 1/88, p. 72.
"A West German firm shipped nuclear fuel. . ." *New Scientist,* 1/7/89, p. 24.

Chapter 18: The Solar/Fusion Fix

"The Israelis have tried. . ." Teller, p. 223.
"Ten years ago a small-scale. . ." *Science,* 8/15/80.
"Photovoltaic cells. . ." Hubbard.
"In England they're working on. . ." *New Scientist,* 1/15/87.
"What could be safer. . ." *New Scientist,* 5/18/78.
"Indeed, the government thought so. . ." Teller, p. 203.
"The other solution being chased. . ." *Science,* 5/16/86; *Physics Today,* 1/88 and 1/89.
"In 1989 the news journal. . ." *Physics Today,* 1/89, p. S-60.
"In 1975, at a series of lectures. . ." Teller, p. 216.
"In 1986 the Electric. . ." and "In the spring of 1989 the National Academy. . ." *New Scientist,* 5/27/89.

Chapter 19: Bathtub Diplomacy

"A group of researchers at the Universities of Copenhagen. . ." *New Scientist,* 6/24/89.
"A censored version. . ." *Science,* 1/20/89.
"The American Institute of Physics. . ." *Physics Today,* letter from Roy Bishop, 3/89, p. 156.
"The list of the top three industrial recipients. . ." *Science,* 11/11/88.

Chapter 20: The Power and the Glory

"As people have been leaving my store today. . ." Miami *Herald,* 2/3/89.
"Over the years. . ." *Science,* 1/6/89.
"Roger Revelle. . ." This and the following from Revkin (1988).
"Wilt thou go with Me. . ." The Bible, 1st Kings, 22:4.

Final Quotes

Edward Teller, from *Energy.*
Yo-Yo Ma from *Profiles* by David Blum in *The New Yorker,* 5/1/89.

Bibliography

Adams, J. A. S., M. S. M. Mantovani, and L. L. Lundell, "Wood versus Fossil Fuel," in *Science,* 4/1/77.

Anspaugh, L. R., R. J. Catlin, and M. Goldman, "The Global Impact of the Chernobyl Reactor Accident," in *Science,* 12/16/88.

Barth, M. C., and J. G. Titus, *Greenhouse Effect and Sea Level Rise,* Van Nostrand Reinhold, 1984.

Berger, A., J. Imbrie, J. Hays, G. Kugla, and B. Saltzman, *Milankovitch and Climate,* D. Reidel, 1984.

Bernard, H. W., *The Greenhouse Effect,* Ballinger, 1980.

Bishop, J. E., "Global Threat," in the *Wall Street Journal,* 10/24/88.

Bolin, B., B. R. Doos, J. Jager, and R. A. Warrick, *The Greenhouse Effect,* John Wiley, 1986.

Brodeur, P., "Annals of Chemistry: Inert," in *The New Yorker,* 4/7/75.

———, "Annals of Chemistry: In the Face of Doubt," in *The New Yorker,* 6/9/86.

Broecker, W. S., "Climatic Change," in *Science,* 8/8/75.

Chown, M., "Smoking Out the Facts of Nuclear Winter," in *New Scientist,* 12/11/86.

———, "Nuclear War," in *New Scientist,* 1/2/86.

Clarke, R., *The Greenhouse Gases,* UNEP/GEMS, 1987.

———, *The Ozone Layer,* UNEP/GEMS, 1987.

Covey, C., S. H. Schneider, and S. L. Thompson, "Global Atmospheric Effects from a Nuclear War," in *Nature,* 3/1/84.

Craig, H., S. L. Miller, and G. J. Wasserberg, *Isotopic and Cosmic Chemistry,* North-Holland, 1964.

Craig, H., C. C. Chou, J. A. Welhan, C. M. Stevens, and A. Engelkemeir, "Isotopic Composition of Methane in Polar Ice Cores," in *Science,* 12/16/88.

Dotto, L., and H. Schiff, *The Ozone War,* Doubleday, 1978.

Emiliani, C., "Ancient Temperatures," in *Scientific American,* 2/58.

Fergusson, J. E., *Inorganic Chemistry and the Earth,* Pergamon Press, 1982.

Flint, R. F., *The Earth and Its History,* Norton, 1973.

Gribbin, J., *Future Weather and the Greenhouse Effect,* Delta, 1982.

———, *The Hole in the Sky,* Bantam, 1988.

Hallion, R. P., "Supersonic Flight," Macmillan, 1972.

Hoyle, F., *Ice,* Continuum, 1981.

———, and G. Hoyle, *Commonsense in Nuclear Energy,* W. H. Freeman, 1980.

Hubbard, H. M., "Photovoltaics Today and Tomorrow," in *Science,* 4/21/89.

Inhaber, H., "Is Solar Power More Dangerous than Nuclear?" in *New Scientist,* 5/18/78.

Kondratyev, K. Ya., *Climate Shocks,* John Wiley, 1988.

Krauskopf, K., *Introduction to Geochemistry, 2nd ed.,* McGraw-Hill, 1979.

Levi, B. G., and T. Rothman, "Nuclear Winter," in *Physics Today,* 9/85.

Levin, H. L., *The Earth Through Time,* Saunders, 1978.

Littmann, M., and D. K. Yeomans, *Comet Halley,* American Chemical Society, 1985.

Lovins, A. B., and L. H. Lovins, "The Avoidable Oil Crisis," in *Atlantic,* 12/87.

McGuirk, J. P., "Planetary Scale Forcing of the January 1977 Weather," in *Science,* 1/20/78.

McLean, D. M., "A Terminal Mesozoic Greenhouse," in *Science,* 8/4/78.

Mintzer, I. M., *A Matter of Degrees,* World Resources Institute, 1987.

Moastersky, R., "Has the Greenhouse Taken Effect?" in *Science News,* 4/30/88.

Nelson, J., "Fractality of Sooty Smoke," *Nature,* 6/22/89.

Nero, A.V., Jr., *A Guidebook to Nuclear Reactors,* University of California Press, 1979.

Patterson, W. C., *The Plutonium Business,* Sierra Club Books, 1984.

Plass, G. N., "The Carbon Dioxide Theory of Climatic Change," in *Tellus,* 8, 2, 1956.

Posey, C., "On the Uranium Trail," in *Atlantic,* 6/84.

Powers, T., "Nuclear Winter and Nuclear Strategy," in *Atlantic,* 11/84.

Raloff, J., "Energy Efficiency," in *Science News,* 5/7/88.

———, "CO_2: How Will We Spell Relief?," in *Science News,* 12/24/88.

Revkin, A. C., "Endless Summer," in *Discover,* 10/88.

———, "Cooling Off the Greenhouse," in *Discover,* 1/89.

Schneider, S. H., "Whatever Happened to Nuclear Winter?," in *Climatic Change,* vol. 12, pp. 215–219, 1988.

———, and L. E. Mesirow, *Genesis Strategy,* Plenum, 1976.

Shell, E. R., "Weather versus Chemicals," in *Atlantic,* 5/87.

Smoluchowski, R., *The Solar System,* Scientific American Library, 1983.

Snell, V., in *The Social Impact of the Chernobyl Disaster,* by David Marples, Macmillan, 1988.

Sparks, B., "Scandal of Nuclear Winter," in *National Review,* 11/15/85.

Stommel H. and E. Stommel, "The Year Without a Summer," *Scientific American,* 6/69.

Tasini, J., "Nuclear Missions," in *Atlantic,* 1/88.

Taylor, J. J., "Improved and Safer Nuclear Power," in *Science,* 4/21/89.

Teller, E., *Energy,* W. H. Freeman, 1979.

Thompson, S. L., and S. H. Schneider, "Nuclear Winter Reappraised," in *Foreign Affairs,* Summer, 1986.

Turco, R. P., O. B. Toon, T. P. Ackerman, J. B. Pollack, and C. Sagan, "Climatic Effects of Nuclear War," in *Scientific American,* 8/84.

Weinberg, A. M., "Are Breeder Reactors Still Necessary," in *Science,* 5/9/86.

Wilson, R., "A Visit to Chernobyl" and "Response," in *Science,* 6/26/87 and 10/2/87.

Woods, C., "Life Without a Sunscreen," in *New Scientist,* 12/10/88.

Index

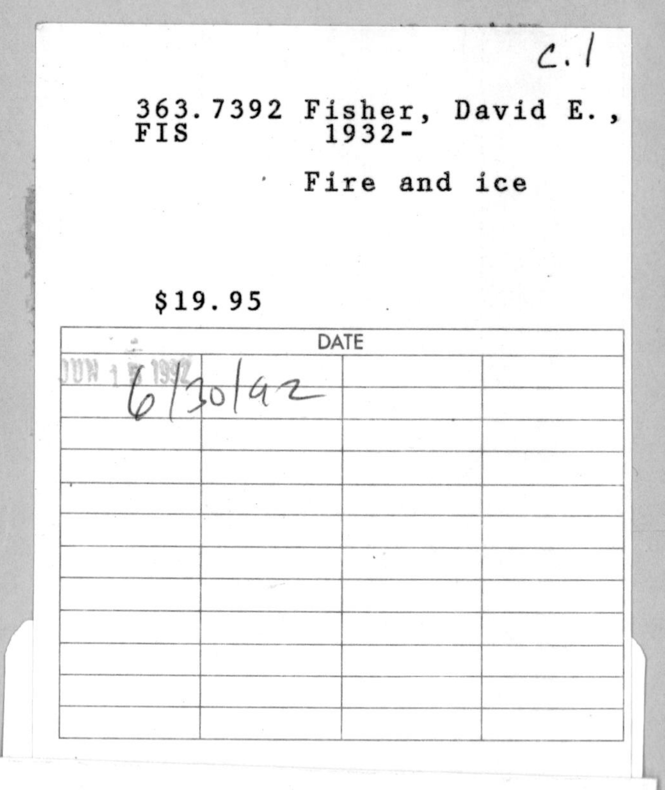